Problems and Solutions in Many-Body Quantum Theory

This practical guidebook provides a comprehensive set of exercises which illustrate the most relevant concepts in a first course on quantum many-body theory as part of advanced undergraduate or graduate courses. The problems come with detailed solutions which can easily be followed either by professors or students. Quantum many-body theory is relevant in several fields, from condensed matter to astrophysics. The problems proposed in this book cover this variety of topics and are illustrated whenever possible with state-of-the-art examples.

Key Features:

- Provides a problems-based approach to quantum many-body theory, unlike existing textbooks.
- In-depth solutions to problems are presented, with the aim to maximize understanding and improve the teaching experience of the subject.
- The multidisciplinary nature of quantum many-body theory is explored in problems that deal with nuclear physics to ultracold atoms and astrophysics.

W0234993

Problems and Solutions in Many-Body Quantum Theory

Sailing the Fermi Sea

Bruno Juliá Díaz, Arnau Rios Huguet,
Héctor Briongos Merino, Javier Rozalén
Sarmiento and Artur Polls Martí

CRC Press
Taylor & Francis Group
Boca Raton London New York

CRC Press is an imprint of the
Taylor & Francis Group, an **informa** business

Designed cover image: Shutterstock

First edition published 2026
by CRC Press
2385 NW Executive Center Drive, Suite 320, Boca Raton FL 33431

and by CRC Press
4 Park Square, Milton Park, Abingdon, Oxon, OX14 4RN

CRC Press is an imprint of Taylor & Francis Group, LLC

© 2026 Bruno Juliá Díaz, Arnau Rios Huguet, Héctor Briongos Merino, Javier Rozalén Sarmiento and Artur Polls Martí

Reasonable efforts have been made to publish reliable data and information, but the author and publisher cannot assume responsibility for the validity of all materials or the consequences of their use. The authors and publishers have attempted to trace the copyright holders of all material reproduced in this publication and apologize to copyright holders if permission to publish in this form has not been obtained. If any copyright material has not been acknowledged please write and let us know so we may rectify in any future reprint.

Except as permitted under U.S. Copyright Law, no part of this book may be reprinted, reproduced, transmitted, or utilized in any form by any electronic, mechanical, or other means, now known or hereafter invented, including photocopying, microfilming, and recording, or in any information storage or retrieval system, without written permission from the publishers.

For permission to photocopy or use material electronically from this work, access www.copyright.com or contact the Copyright Clearance Center, Inc. (CCC), 222 Rosewood Drive, Danvers, MA 01923, 978-750-8400. For works that are not available on CCC please contact mpkbookspermissions@tandf.co.uk

Trademark notice: Product or corporate names may be trademarks or registered trademarks and are used only for identification and explanation without intent to infringe.

ISBN: 978-1-032-04296-1 (hbk)
ISBN: 978-1-032-01355-8 (pbk)
ISBN: 978-1-003-19133-9 (ebk)

DOI: 10.1201/9781003191339

Typeset in Nimbus Roman
by KnowledgeWorks Global Ltd.

Publisher's note: This book has been prepared from camera-ready copy provided by the authors.

Dedication

_____

To Artur Polls Martí, in memoriam

Contents

SECTION I Basics

SECTION II Applications

Preface

This book is the result of the teaching experience associated to a "Many-Body Quantum Theory" module which has existed as part of the Physics curriculum at the University of Barcelona, in different shapes and forms, in the last few decades. The module was mostly carried by Prof. Artur Polls for over 20 years, together with Profs. Manuel Barranco, Muntsa Guilleumas, Bruno Juliá-Díaz, and Àngels Ramos. Prof. Polls was a pioneer of the field with a broad national and international reputation. He passed away in 2020, but this book remains as part of his legacy to a discipline, quantum many-body physics, that he enjoyed. He also taught it with exquisite clarity and, perhaps more importantly, set the foundations for several generations of physicists working on it now.

One of us (Prof. Juliá-Díaz) took over the module in 2020 and curated the last version of the problem sets that are taught in conjunction with standard theory lectures. Dr. Rios joined the project in 2021, but it wasn't until the arrival of PhD students Héctor Briongos Merino and Javier Rozalén Sarmiento that the final book write-up really took off in 2024.

Just like any other University module, the choice of topics in the "Quantum Many-Body Physics" course is biased by the lecturers that teach it. In this case, the broad scope of the quantum many-body physics field is showcased in the problems that verse on nuclear physics; ultracold atoms (including fermions and bosons); and systems with low dimensionality.

A number of excellent books discussing quantum many-body physics have been published over the years with a variety of perspectives [9, 3, 11, 7, 6, 5, 1, 12, 4]. These manuals, typically aimed at master's and graduate levels, often incorporate problems as work-at-home examples. Unfortunately, they rarely include the actual solutions to the problems, potentially causing a gap in understanding. In some cases, this can hamper the learning activity of students that cannot solve them for whatever reason.

The aim of our book is precisely to fill this gap in the current literature. We have prepared a manual of problems and solutions for master's and graduate students that want to deepen their practical understanding of quantum many-body physics. Because many theory manuals are already available in the literature, our approach has focused mostly on the solutions themselves. We provide a very cursory reminder of theory essentials in each chapter, which focuses on specific content, but the heart of our book is the solution to the proposed problems. We have also prepared two appendices discussing briefly second quantization (Appendix A) and Wick's theorem (Appendix B), to furnish some further context to problems that required so.

In our minds, the target audience of this book are master's or graduate students trying to start their way into the beautiful, yet sometimes challenging, discipline that is quantum many-body physics. All the problems in this collection have analytical solutions. As many-body practitioners know, analytical solutions are the exception in this field, not the rule. Having said that, we also feel that in laying the basic groundwork for the discipline, a solid understanding of some concepts cannot be achieved without some of the analytical work we dissect here.

Some of the problems we propose are classics in the field: computing the pair distribution function of the free Fermi gas or an explicit calculation of the Lindhard function. Others, like those on Wick's theorem, can be daunting to work on at first, but provide a solid understanding that is of course also relevant for other disciplines, such as quantum field theory. Finally, we give some examples that are not mathematically cumbersome, but we hope that they yield some intuition of the physics at play.

In most cases, we have provided a step-by-step solution to the problems, including mathematical details that may have been skipped in the interest of brevity. We hope that, removing the intermediate mathematical complexity, may render a more physically-oriented learning experience. In doing so, we hope to convey Prof. Polls' teaching style, which was very much one of being extremely careful in exposition; elucidating all possible pitfalls and errors; and ultimately reaching the correct solution, providing along the way a clear physical intuition.

Contributors

Bruno Juliá Díaz
Departament de Física Quàntica i
  Astrofísica and Institut de Ciències
  del Cosmos (ICCUB), Universitat de
  Barcelona
Barcelona, Spain

Artur Polls Martí
Departament de Física Quàntica i
  Astrofísica and Institut de Ciències
  del Cosmos (ICCUB), Universitat de
  Barcelona
Barcelona, Spain

Arnau Rios Huguet
Departament de Física Quàntica i
  Astrofísica and Institut de Ciències
  del Cosmos (ICCUB), Universitat de
  Barcelona and University of Surrey,
  Guildford (UK)

Héctor Briongos Merino
Departament de Física Quàntica i
  Astrofísica and Institut de Ciències
  del Cosmos (ICCUB), Universitat de
  Barcelona
Barcelona, Spain

Javier Rozalén Sarmiento
Departament de Física Quàntica i
  Astrofísica and Institut de Ciències
  del Cosmos (ICCUB), Universitat de
  Barcelona
Barcelona, Spain

Section I

Basics

1 Symmetries

Symmetry considerations lie at the very heart of quantum many-body theory. In particular, symmetries dictate and restrict the properties of many-body wavefunctions. A many-body system with N particles can be described at zero temperature by a many-body wavefunction $\Psi(\vec{x}_1, \cdots, \vec{x}_A)$, where $\vec{x}_i$ represents the coordinate of particle i. Note that this coordinate can be discrete, e.g. a lattice site in a spin system; continuous, if, for instance, it describes the position of a particle; or a combination of both, when we work with particles of definite position and spin, $\vec{x}_i = (\vec{x}_i, \sigma_i)$.

Permutation symmetry dictates how the many-body wavefunction of a system transforms when two particles are exchanged. Consider the permutation operator P_{ij}, which exchanges the coordinates of particles i and j in the many-body wavefunction,

$$P_{ij}\Psi(\vec{x}_1, \vec{x}_2, \ldots, \vec{x}_i, \ldots, \vec{x}_j, \ldots, \vec{x}_N) = \Psi(\vec{x}_1, \vec{x}_2, \ldots, \vec{x}_j, \ldots, \vec{x}_i, \ldots, \vec{x}_N). \qquad (1.0.1)$$

Since applying the operator P_{ij} twice leads to the original wavefunction, we know that $P_{ij}^2 = 1$. In consequence, the eigenvalues of this operator when acting on a wavefunction can either be $+1$ or -1.

The *symmetrization postulate* indicates that the only admissible many-body wavefunctions are either fully symmetric or fully antisymmetric with respect to all permutation operators. This is to be expected because we assume that the system is formed of A indistinguishable particles. Permuting any one particle with another cannot change the state of the system more than by a single phase. This phase, or the sign associated to a permutation of coordinates, is one of the more fundamental properties of a many-body wavefunction. It takes a value of $+1$ for bosonic systems, or a value of -1 for fermionic systems.

In this chapter, we provide some insight on the symmetrization and antisymmetrization operators, two key projectors for the two fundamental species of identical particles, and their effect on wavefunctions of quantum many-body systems.

PROBLEM 1.1
Symmetrization and antisymmetrization operators for $N = 2$

The symmetrization and antisymmetrization operators for 2 identical particles are defined as:

$$\hat{S} = \frac{1}{2}\left(1 + \hat{P}_{12}\right), \qquad \hat{A} = \frac{1}{2}\left(1 - \hat{P}_{12}\right). \qquad (1.1.1)$$

Here, $\hat{P}_{12}$ is the exchange operators of particles 1 and 2. Show explicitly that:

(a) $\hat{S}^2 = \hat{S}$ and $\hat{A}^2 = \hat{A}$;

(b) $\hat{S} + \hat{A} = 1$;

(c) $\hat{S}\hat{A} = \hat{A}\hat{S} = 0$;

(d) $\hat{P}_{12}\hat{S} = \hat{S}$, and $\hat{P}_{12}\hat{A} = -\hat{A}$.

(a) We start with the symmetrization operator. We apply it twice and find, using $P_{12}P_{12} = 1$ where 1 is the identity operator,

$$\hat{S}\hat{S} = \frac{1}{2}\left(1+\hat{P}_{12}\right)\frac{1}{2}\left(1+\hat{P}_{12}\right) = \frac{1}{4}\left(1+\hat{P}_{12}+\hat{P}_{12}+1\right) = \frac{1}{2}\left(1+\hat{P}_{12}\right) = \hat{S}.$$

Proceeding similarly with the antisymmetrization operator:

$$\hat{A}\hat{A} = \frac{1}{2}\frac{1}{2}\left(1-\hat{P}_{12}\right)\left(1-\hat{P}_{12}\right) = \frac{1}{4}\left(1-\hat{P}_{12}-\hat{P}_{12}+1\right) = \frac{1}{2}\left(1-\hat{P}_{12}\right) = \hat{A}.$$

We have shown that both operators are idempotent, i.e., they fulfill the condition $\hat{O}^2 = \hat{O}$. This is a sufficient condition to say that both are projectors.

(b) The sum of the two operators is easily identified with the identity,

$$\hat{S}+\hat{A} = \frac{1}{2}\left(1+\hat{P}_{12}\right) + \frac{1}{2}\left(1-\hat{P}_{12}\right) = \frac{1}{2}\left(1+\hat{P}_{12}+1-\hat{P}_{12}\right) = 1.$$

This property shows that operators $\hat{S}$ and $\hat{A}$ form a set of operators that covers all Hilbert space of 2 particles, and thus each state can be decomposed in a symmetric and an antisymmetric part by means of these projectors.

(c) In contrast, the product of the two operators is null:

$$\hat{S}\hat{A} = \frac{1}{2}\left(1+\hat{P}_{12}\right)\frac{1}{2}\left(1-\hat{P}_{12}\right) = \frac{1}{4}\left(1+\hat{P}_{12}-\hat{P}_{12}-1\right) = 0.$$

(d) We now demonstrate that a permutation of two particles leaves the symmetrization operator unchanged:

$$\hat{P}_{12}\hat{S} = \hat{P}_{12}\frac{1}{2}\left(1+\hat{P}_{12}\right) = \frac{1}{2}\left(\hat{P}_{12}+1\right) = \hat{S}.$$

Exchanging two particles before the antisymmetrization operator instead brings in a minus sign:

$$\hat{P}_{12}\hat{A} = \hat{P}_{12}\frac{1}{2}\left(1-\hat{P}_{12}\right) = \frac{1}{2}\left(\hat{P}_{12}-1\right) = -\hat{A}.$$

PROBLEM 1.2
Symmetrization and antisymmetrization operators.

The definitions for the symmetrization

$$\hat{S} = \frac{1}{N!} \sum_{p \in S_N} \hat{P}, \tag{1.2.1}$$

and antisymmetrization operator

$$\hat{A} = \frac{1}{N!} \sum_{p \in S_N} (-1)^{\varepsilon(p)} \hat{P}, \tag{1.2.2}$$

for N identical particles include the permutation operator, $\hat{P}$, running over all permutations of the symmetric group of N objects. $\varepsilon(p)$ is the signature of a given permutation, p. Using these definitions, show that:

(a) $\hat{S}^2 = \hat{S}$ and $\hat{A}^2 = \hat{A}$;

(b) $\hat{S}\hat{A} = \hat{A}\hat{S} = 0$;

(c) $\hat{P}_{jk}\hat{S} = \hat{S}$; and

(d) $\hat{P}_{jk}\hat{A} = -\hat{A}$.

(a) We start by applying the symmetrization operator twice to get $\hat{S}^2$. We use two different dummy operators, $\hat{P}$ and $\hat{Q}$, for the corresponding sums and get:

$$\hat{S}^2 = \hat{S}\hat{S} = \left(\frac{1}{N!} \sum_{P \in S_N} \hat{P} \right) \left(\frac{1}{N!} \sum_{Q \in S_N} \hat{Q} \right) = \frac{1}{N!} \frac{1}{N!} \sum_{Q \in S_N} \sum_{P \in S_N} \hat{P}\hat{Q}.$$

When $\hat{P}$ runs over all S_N, due to the closure property of groups, $\hat{P}\hat{Q}$ (where $\hat{Q}$ is a given element of S_N) runs over all S_N too. In consequence, we can replace $\hat{P}\hat{Q}$ by a single permutation operator, $\hat{R}$. The sum over $\hat{Q}$ becomes redundant, so that

$$\hat{S}^2 = \frac{1}{N!} \frac{1}{N!} \sum_{Q \in S_N} \sum_{R \in S_N} \hat{R} = \frac{1}{N!} \frac{1}{N!} N! \sum_{R \in S_N} \hat{R} = \hat{S}.$$

We follow a similar strategy to compute $\hat{A}^2$, following the same steps and introducing a new $\hat{R}$ operator:

$$\hat{A}^2 = \hat{A}\hat{A} = \left(\frac{1}{N!} \sum_{Q \in S_N} (-1)^{\varepsilon(\hat{Q})} \hat{Q} \right) \left(\frac{1}{N!} \sum_{P \in S_N} (-1)^{\varepsilon(P)} \hat{P} \right)$$

$$= \frac{1}{N!} \frac{1}{N!} \sum_{Q \in S_N} \sum_{P \in S_N} (-1)^{\varepsilon(Q) + \varepsilon(P)} \hat{Q}\hat{P} = \frac{1}{N!} \frac{1}{N!} \sum_{Q \in S_N} \sum_{R \in S_N} (-1)^{\varepsilon(R)} \hat{R}$$

$$= \frac{1}{N!} \frac{1}{N!} N! \sum_{R \in S_N} (-1)^{\varepsilon(R)} \hat{R} = \hat{A},$$

where we have used the following property:

$$(-1)^{\varepsilon(Q) + \varepsilon(P)} = (-1)^{\varepsilon(R)},$$

which holds given that the parity of a given permutation is $(-1)^n$ with n the smallest number of 2-particle transpositions needed to obtain the permutation.

We have just shown that both the symmetrization and the antisymmetrization operators, $\hat{S}$ and $\hat{A}$, are idempotent, and this condition is sufficient to demonstrate that they are projectors.

(b) To prove the equality $\hat{S}\hat{A} = 0$, we write explicitly the sums associated to each operator in terms of the permutation operators,

$$\hat{S}\hat{A} = \frac{1}{N!} \frac{1}{N!} \sum_{Q \in S_N} \hat{Q} \sum_{P \in S_N} (-1)^{\varepsilon(P)} \hat{P} = \frac{1}{N!} \frac{1}{N!} \sum_{Q \in S_N} (-1)^{\varepsilon(Q)} \sum_{P \in S_N} (-1)^{\varepsilon(Q) + \varepsilon(P)} \hat{Q}\hat{P}.$$

In this last line, we have used the trick $(-1)^{\varepsilon(Q)} (-1)^{\varepsilon(Q)} = 1$. We now proceed by noting, again, that the product of two permutations, $\hat{Q}\hat{P}$, is a permutation $\hat{R}$, and find:

$$\hat{S}\hat{A} = \frac{1}{N!} \sum_{Q \in S_N} (-1)^{\varepsilon(Q)} \frac{1}{N!} \sum_{R \in S_N} (-1)^{\varepsilon(R)} \hat{R} = \frac{1}{N!} \sum_{Q \in S_N} (-1)^{\varepsilon(Q)} \hat{A}$$

$$= \hat{A} \frac{1}{N!} \sum_{Q \in S_N} (-1)^{\varepsilon(Q)} = 0.$$

(c) We now consider the exchange of two particles, j and k, in front of the symmetrization operator. Using explicitly Eq. (1.2.1), we find:

$$\hat{P}_{jk} \hat{S} = \hat{P}_{jk} \frac{1}{N!} \sum_{P \in S_N} \hat{P} = \frac{1}{N!} \sum_{P \in S_N} \hat{P}_{jk} \hat{P} = \frac{1}{N!} \sum_{R \in S_N} \hat{R} = \hat{S}.$$

(d) A very similar derivation, but starting from Eq. (1.2.2), yields:

$$\hat{P}_{jk} \hat{A} = \hat{P}_{jk} \frac{1}{N!} \sum_{P \in S_N} (-1)^{\varepsilon(P)} \hat{P} = \frac{1}{N!} \sum_{P \in S_N} (-1)^{\varepsilon(P_{jk})} (-1)^{\varepsilon(P_{jk})} (-1)^{\varepsilon(P)} \hat{P}_{jk} \hat{P}$$

$$= \frac{1}{N!} (-1)^{\varepsilon(P_{jk})} \sum_{P \in S_N} (-1)^{\varepsilon(P_{jk}) + \varepsilon(P)} \hat{P}_{jk} \hat{P} = -\frac{1}{N!} \sum_{R \in S_N} (-1)^{\varepsilon(R)} \hat{R} = -\hat{A},$$

where we have taken into account that

$$(-1)^{\varepsilon(P_{jk})} = -1,$$

and that

$$(-1)^{\varepsilon(P_{jk})+\varepsilon(P)} = (-1)^{\varepsilon(P_{jk}P)}.$$

PROBLEM 1.3
Symmetrization in the space representation

Consider the properly symmetrized wavefunction of two spinless bosons, which are distributed over two single-particle levels, 1 and 2,

$$\Psi(\vec{r}_1,\vec{r}_2) = \frac{1}{\sqrt{2}}\left[\psi_1(\vec{r}_1)\psi_2(\vec{r}_2) + \psi_1(\vec{r}_2)\psi_2(\vec{r}_1)\right]. \tag{1.3.1}$$

The two single-particle states are stationary and properly normalized. Assuming that state 1 has positive parity and state 2 has negative parity:

(a) Determine the probability of finding a particle at position $\vec{r}$, $P(\vec{r})$, if the position of the other one is arbitrary.

(b) Find the probability of finding one particle in the upper half-space, $z \geq 0$.

(c) Find the probability of finding two particles in the upper half-space, $z \geq 0$.

(d) Answer these questions for two fermions in the same states.

(e) If the two particles were distinguishable, we would know in which state each particle sits. The wavefunction would be a simple product state, $\Psi(\vec{r}_1,\vec{r}_2) = \psi_1(\vec{r}_1)\psi_2(\vec{r}_2)$. Answer the previous questions for this case.

This problem is inspired by problem 15.5 in Ref. [15].

(a) We start by finding the probability of particle 1 without knowing anything about the position of particle 2. We marginalize over one of the positions by integrating over the whole volume:

$$P(\vec{r}) = \int_{-\infty}^{\infty} d\vec{r}_2 |\Psi(\vec{r}_1 = \vec{r}, \vec{r}_2)|^2$$

$$= \frac{1}{2} |\psi_1(\vec{r})|^2 \underbrace{\int_{-\infty}^{\infty} d\vec{r}_2 |\psi_2(\vec{r}_2)|^2}_{=1} + \frac{1}{2} \psi_1^*(\vec{r}) \psi_2(\vec{r}) \underbrace{\int_{-\infty}^{\infty} d\vec{r}_2 \psi_2^*(\vec{r}_2) \psi_1(\vec{r}_2)}_{=0} +$$

$$+ \frac{1}{2} \psi_2^*(\vec{r}) \psi_1(\vec{r}) \underbrace{\int_{-\infty}^{\infty} d\vec{r}_2 \psi_1^*(\vec{r}_2) \psi_2(\vec{r}_2)}_{=0} + \frac{1}{2} |\psi_2(\vec{r})|^2 \underbrace{\int_{-\infty}^{\infty} d\vec{r}_2 |\psi_1(\vec{r}_2)|^2}_{=1}$$

$$= \frac{1}{2} |\psi_1(\vec{r})|^2 + \frac{1}{2} |\psi_2(\vec{r})|^2 . \qquad (1.3.2)$$

The integrals of the square wavefunction are assumed to be normalized. The crossed terms are instead zero due to parity. To see this, we use the parity of each state, $\psi_1(-\vec{r}) = +\psi_1(\vec{r})$ but $\psi_2(-\vec{r}) = -\psi_2(\vec{r})$, and look at the crossed terms under a change of the integration variable $\vec{r}_2 \to -\vec{r}_2$:

$$\int_{-\infty}^{\infty} d\vec{r}_2 \psi_2^*(\vec{r}_2) \psi_1(\vec{r}_2) = \int_{-\infty}^{\infty} d\vec{r}_2 \psi_2^*(-\vec{r}_2) \psi_1(-\vec{r}_2)$$

$$= -\int_{-\infty}^{\infty} d\vec{r}_2 \psi_2^*(\vec{r}_2) \psi_1(\vec{r}_2) \equiv 0.$$

The result of Eq. 1.3.2 has an "intuitive" explanation, in that, upon integrating over the second particle, we find that the remaining particle is equally distributed between either one of the single-particle wavefunctions.

(b) To determine the probability of finding one particle over the upper half-space, we integrate the previous result for $z \geq 0$,

$$W_{z \geq 0} = \int_{z \geq 0} d\vec{r} P(\vec{r}) = \frac{1}{2} \int_{z \geq 0} d\vec{r} |\psi_1(\vec{r})|^2 + \frac{1}{2} \int_{z \geq 0} d\vec{r} |\psi_2(\vec{r})|^2 .$$

The two wavefunctions have definite parity, so the integrals over the lower and upper half-spaces are the same, as shown below,

$$\int_{z \geq 0} d\vec{r} |\psi_i(\vec{r})|^2 = \int_{-\infty}^{\infty} dx \int_{-\infty}^{\infty} dy \int_0^{\infty} dz |\psi_i(x, y, z)|^2$$

$$= \int_{-\infty}^{\infty} dx \int_{-\infty}^{\infty} dy \int_0^{\infty} dz |\psi_i(x, y, -z)|^2$$

$$= \int_{-\infty}^{\infty} dx \int_{-\infty}^{\infty} dy \int_{-\infty}^0 dz |\psi_i(x, y, z)|^2 = A.$$

We choose the variable A to denote this half-volume integral. Moreover, since the wavefunctions are normalized over the whole volume,

$$\int_{-\infty}^{\infty} d\vec{r} |\psi_i(\vec{r})|^2 = \int_{-\infty}^{\infty} dx \int_{-\infty}^{\infty} dy \int_0^{\infty} dz |\psi_i(x, y, z)|^2$$

$$+ \int_{-\infty}^{\infty} dx \int_{-\infty}^{\infty} dy \int_{-\infty}^0 dz |\psi_i(x, y, z)|^2 = 2A = 1,$$

so clearly $A = \frac{1}{2}$. With this, we find that

$$W_{z \geq 0} = = \frac{1}{2} \times \frac{1}{2} + \frac{1}{2} \times \frac{1}{2} = \frac{1}{2}.$$

In other words, there is a 50% chance of finding one particle in the upper half-space.

(c) To determine the probability of finding two particles in the upper half-space, $z \geq 0$, we need to look at the two-body probability distribution. This is obtained from the two-body wavefunction, integrated over the upper half-space of each particle. This involves known integrals over squared wavefunctions in this region as well as an unknown, possibly complex term involving the crossed integral of the two states in the upper half plane, which we dub I:

$$W_{z_1 \geq 0, z_2 \geq 0} = \int_{z_1 \geq 0} d\vec{r}_1 \int_{z_2 \geq 0} d\vec{r}_2 |\psi(\vec{r}_1, \vec{r}_2)|^2$$

$$= \frac{1}{2} \left[\underbrace{\int_{z_1 \geq 0} d\vec{r}_1 |\psi_1(\vec{r})|^2}_{=1/2} \underbrace{\int_{z_2 \geq 0} d\vec{r}_2 |\psi_2(\vec{r})|^2}_{=1/2} \right.$$

$$+ \underbrace{\int_{z_1 \geq 0} d\vec{r}_1 \psi_1^*(\vec{r}_1) \psi_2(\vec{r}_1)}_{=I} \underbrace{\int_{z_2 \geq 0} d\vec{r}_2 \psi_2^*(\vec{r}_2) \psi_1(\vec{r}_2)}_{=I^*}$$

$$+ \underbrace{\int_{z_1 \geq 0} d\vec{r}_1 \psi_2^*(\vec{r}_1) \psi_1(\vec{r}_1)}_{=I^*} \underbrace{\int_{z_2 \geq 0} d\vec{r}_2 \psi_1^*(\vec{r}_2) \psi_2(\vec{r}_2)}_{=I}$$

$$\left. + \underbrace{\int_{z_1 \geq 0} d\vec{r}_1 |\psi_2(\vec{r})|^2}_{=1/2} \underbrace{\int_{z_2 \geq 0} d\vec{r}_2 |\psi_1(\vec{r}_2)|^2}_{=1/2} \right]$$

$$= \frac{1}{4} + |I|^2.$$

With this, we find that the probability of finding two particles in the upper half-space is either 25% or larger. Indeed, the probability is increased by a real and positive factor, $|I|^2$, which depends on the problem under consideration.

(d) For fermions, the antisymmetrized wavefunction reads

$$\Psi(\vec{r}_1, \vec{r}_2) = \frac{1}{\sqrt{2}} [\psi_1(\vec{r}_1) \psi_2(\vec{r}_2) - \psi_1(\vec{r}_2) \psi_2(\vec{r}_1)], \qquad (1.3.3)$$

and the results for problems (a) and (b) are the same. In (a) the mixed terms, which carry a minus sign for fermions, cancel, and so one obtains the same $P(\vec{r})$. In contrast, for (c), the sign of the mixed terms is relevant, and opposite to the bosonic case. One finds

$$W_{z_1 \geq 0, z_2 \geq 0} = \frac{1}{4} - |I|^2.$$

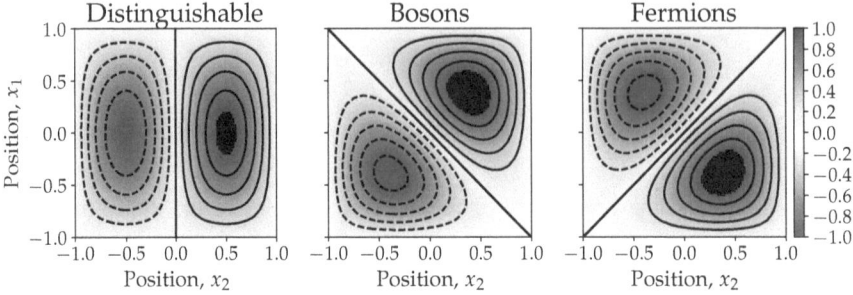

Figure 1.1   Contour density plots for the two-body wavefunction $\Psi(x_1,x_2)$ from different product states using the ground and first excited states of an infinite well with limits at $x = \pm 1$. The left panel shows the product of two distinguishable particle states. The central and right panel correspond to the bosonic and fermionic cases. Note the very different structure of these wavefunctions along the $x_1 = x_2$ line. Dashed (solid) lines represent negative (positive) contours spaced in steps of 0.1.

In other words, the probability of finding 2 fermions on the upper half space is less than 25%. In fact, this is a factor $2|I|^2$ smaller than in the bosonic case.

(e) The results in the distinguishable case are very different. To start with, the one-particle probability depends on which particle we are looking into. The probability of finding particle 1 is obtained from the total two-body wavefunction by marginalizing over particle 2:

$$P_1(\vec{r}) = \int_{-\infty}^{\infty} d\vec{r}_2 |\Psi(\vec{r} = \vec{r}_1, \vec{r}_2)|^2 = |\psi_1(\vec{r})|^2 \underbrace{\int_{-\infty}^{\infty} d\vec{r}_2 |\psi_2(\vec{r}_2)|^2}_{=1} = |\psi_1(\vec{r})|^2,$$

which yields the single-particle probability of particle 1. In contrast, for particle 2, the result is just the probability of particle 2,

$$P_2(\vec{r}) = \int_{-\infty}^{\infty} d\vec{r}_1 |\Psi(\vec{r} = \vec{r}_1, \vec{r}_2)|^2 = \underbrace{\int_{-\infty}^{\infty} d\vec{r}_1 |\psi_1(\vec{r})|^2}_{=1} |\psi_2(\vec{r}_2)|^2 = |\psi_2(\vec{r})|^2.$$

In addition, the probability to find either particle in the upper half-space is just $1/2$. The probability to find the two of them simultaneously in this same region is $1/4$.

As a practical example of the importance of the symmetrization of wavefunctions, we show in Fig. 1.1 the result of three different symmetrizations of a two-body wavefunction. We now have an infinite square well Hamiltonian, with bounds placed at $x = \pm L$. The single-particle wavefunctions for this case are:

$$\psi_n(x) = \begin{cases} \sqrt{\frac{1}{L}} \cos\left(\frac{n}{2}\frac{\pi}{L}x\right) & \text{if } n \text{ odd}, \\ \sqrt{\frac{1}{L}} \sin\left(\frac{n}{2}\frac{\pi}{L}x\right) & \text{if } n \text{ even}. \end{cases} \tag{1.3.4}$$

For plotting purposes, we work with $L = 1$. We use the convention where $n = 1$ represents the ground state here.

Two distinguishable particles have a simple product wavefunction. We choose in which state each particle is placed, and we put particle 1 in the ground state, ψ_1, and particle 2, in the first excited state, ψ_2:

$$\Psi(x_1, x_2) = \psi_1(x_1)\psi_2(x_2) = \cos\left(\frac{\pi}{2}x_1\right)\sin(\pi x_2) . \qquad (1.3.5)$$

A contour plot corresponding to this two-body wavefunction is shown in the left panel of Fig. 1.1. The wavefunction is only zero if $x_2 = 0$.

For two bosons in the same two states, one finds the two-body wavefunction:

$$\begin{aligned}
\Psi(x_1, x_2) &= \frac{1}{\sqrt{2}}[\psi_1(x_1)\psi_2(x_2) + \psi_1(x_2)\psi_2(x_1)] \\
&= \frac{1}{\sqrt{2}}\left[\cos\left(\frac{\pi}{2}x_1\right)\sin(\pi x_2) + \sin(\pi x_1)\cos\left(\frac{\pi}{2}x_2\right)\right].
\end{aligned} \qquad (1.3.6)$$

The wavefunction only cancels when $x_1 = -x_2$. This is shown in the central panel of Fig. 1.1. We stress that two bosons could also sit on the same state, with a total wavefunction $\Psi(x_1, x_2) = \psi_1(x_1)\psi_1(x_2)$. This product state may have a lower energy.

In contrast, for two fermions, the wavefunction has a $-$ sign and thus

$$\begin{aligned}
\Psi(x_1, x_2) &= \frac{1}{\sqrt{2}}[\psi_1(x_1)\psi_2(x_2) - \psi_1(x_2)\psi_2(x_1)] \\
&= \frac{1}{\sqrt{2}}\left[\cos\left(\frac{\pi}{2}x_1\right)\sin(\pi x_2) - \sin(\pi x_1)\cos\left(\frac{\pi}{2}x_2\right)\right].
\end{aligned} \qquad (1.3.7)$$

Here, the nodal line occurs when $x_1 = x_2$, as displayed in the right panel of Fig. 1.1.

2 Second quantization

In non-relativistic quantum mechanics, each particle is characterized by a set of parameters. These are intrinsic to a given particle, and they are the same for a given particle species. Think, for instance, of the mass, the charge, or the spin of an electron. All electrons have the same mass, charge and spin and, from these properties alone, we cannot distinguish two different electrons. Based on these properties, we can divide particles into classes depending on their intrinsic parameters. Particles belonging to the same class are said to be *indistinguishable*.

The way this works is different in quantum than in classical mechanics. In classical mechanics, we can always distinguish particles depending on some of their properties, like their positions or momenta. The positions of two identical quantum mechanical particles, in contrast, cannot be distinguished. In quantum mechanics, we face a different problem with a different meaning for "indistinguishability" than in classical mechanics.

The natural language to describe indistinguishable particles in quantum mechanics is the so-called second quantization formalism. Rather than working in a Hilbert space with a fixed number of particles, $\mathscr{H}(N)$, in second quantization we let go of the idea of a fixed number of particles, and form a Hilbert space that is large enough to accommodate an undetermined number of particles, called the Fock space. The mathematical tools used in this formalism also differ from those found in first quantization. We refer the reader to Appendix A for a more detailed introduction to the second quantization formalism.

In this chapter, we explore the basics of second quantization through different problems. These exercises take us from the most basic properties of creation and destruction operators, to the operators we can build with them, and to more advanced constructions such as field and spin operators.

PROBLEM 2.1
Fock space basis states are eigenvectors of the particle number operator $\hat{N}$

Using the anticommutation relations for the creation and destruction operators in a system of identical fermions, show that:

(a) $\hat{a}_\alpha^\dagger |0\rangle$ is an eigenstate of the particle number operator $\hat{N} = \sum_\alpha \hat{a}_\alpha^\dagger \hat{a}_\alpha$.

(b) $\hat{a}_\beta^\dagger \hat{a}_\gamma^\dagger |0\rangle$ is also an eigenstate of $\hat{N}$.

In both cases, determine the corresponding eigenvalues.

(a) The trick to this problem is to consider first the state $\hat{a}_\alpha^\dagger |0\rangle$ and then apply the number operator to it on the left. We are told these are fermions, so we can then immediately use anticommutation relations to find:

$$\hat{N}\left(\hat{a}_\alpha^\dagger |0\rangle\right) = \left[\sum_\beta \hat{a}_\beta^\dagger \hat{a}_\beta\right]\hat{a}_\alpha^\dagger |0\rangle = \sum_\beta \hat{a}_\beta^\dagger \underbrace{\hat{a}_\beta \hat{a}_\alpha^\dagger}_{\delta_{\alpha\beta}-\hat{a}_\alpha^\dagger \hat{a}_\beta} |0\rangle$$

$$= \sum_\beta \hat{a}_\beta^\dagger \left(\delta_{\alpha\beta} - \hat{a}_\alpha^\dagger \hat{a}_\beta\right)|0\rangle = \hat{a}_\alpha^\dagger |0\rangle\,,$$

where in the last equality we employ the relation $\hat{a}_\beta |0\rangle = 0$. In other words, the state $|\alpha\rangle = \hat{a}_\alpha^\dagger |0\rangle$ is an eigenvector of the particle number operator with eigenvalue 1,

$$\hat{N}\hat{a}_\alpha^\dagger |0\rangle = 1\, \hat{a}_\alpha^\dagger |0\rangle\,.$$

(b) For the state $\hat{a}_\beta^\dagger \hat{a}_\alpha^\dagger |0\rangle$, the situation is very similar. We apply the number operator to the left of the state and employ the anticommutation relations twice:

$$\hat{N}(\hat{a}_\beta^\dagger \hat{a}_\alpha^\dagger |0\rangle) = \left[\sum_\gamma \hat{a}_\gamma^\dagger \hat{a}_\gamma\right]\hat{a}_\beta^\dagger \hat{a}_\alpha^\dagger |0\rangle = \sum_\gamma \hat{a}_\gamma^\dagger (\delta_{\gamma\beta} - \hat{a}_\beta^\dagger \hat{a}_\gamma)\hat{a}_\alpha^\dagger |0\rangle$$

$$= \hat{a}_\beta^\dagger \hat{a}_\alpha^\dagger |0\rangle - \sum_\gamma \hat{a}_\gamma^\dagger \hat{a}_\beta^\dagger (\delta_{\gamma\alpha} - \hat{a}_\alpha^\dagger \hat{a}_\gamma)|0\rangle$$

$$= \hat{a}_\beta^\dagger \hat{a}_\alpha^\dagger |0\rangle - \hat{a}_\alpha^\dagger \hat{a}_\beta^\dagger |0\rangle = 2\,\hat{a}_\alpha^\dagger \hat{a}_\beta^\dagger |0\rangle\,.$$

Therefore, the state $\hat{a}_\beta^\dagger \hat{a}_\alpha^\dagger |0\rangle$ is indeed an eigenvector of $\hat{N}$ with eigenvalue 2.

PROBLEM 2.2
The particle number operator $\hat{N}$ commutes with the Hamiltonian

Show that the particles number operator, $\hat{N} = \sum_\alpha \hat{a}_\alpha^\dagger \hat{a}_\alpha$, commutes with the Hamiltonian, $\hat{H}$, of a system of fermions. To this end, consider the partition $\hat{H} = \hat{H}_0 + \hat{H}_1$, with

(a) A one-body operator term,

$$\hat{H}_0 = \sum_{\alpha\beta} t_{\alpha\beta} \hat{a}_\alpha^\dagger \hat{a}_\beta\,,$$

where the matrix elements are defined as $t_{\alpha\beta} = \langle\alpha|\hat{T}|\beta\rangle$.

(b) A two-body operator term,

$$\hat{H}_1 = \frac{1}{2}\sum_{\alpha\beta\delta\gamma} v_{\alpha\beta\gamma\delta} \hat{a}_\alpha^\dagger \hat{a}_\beta^\dagger \hat{a}_\delta \hat{a}_\gamma\,,$$

with matrix elements $v_{\alpha\beta\gamma\delta} = \langle\alpha\beta|\hat{V}|\gamma\delta\rangle$.

(c) How does the result change for bosons?

(a) We look first at the commutator of the one-body operator with the number operator, $[\hat{H}_0, \hat{N}] = \sum_{\alpha\beta} t_{\alpha\beta} [\hat{a}_\alpha^\dagger \hat{a}_\beta, \hat{N}]$. We compute each term within the sum separately. We employ the following relation,

$$[\hat{a}_\alpha^\dagger \hat{a}_\beta, \hat{N}] = \hat{a}_\alpha^\dagger \hat{a}_\beta \hat{N} - \hat{N}\hat{a}_\alpha^\dagger \hat{a}_\beta = \hat{a}_\alpha^\dagger \hat{a}_\beta \hat{N} - \hat{a}_\alpha^\dagger \hat{N}\hat{a}_\beta + \hat{a}_\alpha^\dagger \hat{N}\hat{a}_\beta - \hat{N}\hat{a}_\alpha^\dagger \hat{a}_\beta$$
$$= \hat{a}_\alpha^\dagger [\hat{a}_\beta, \hat{N}] + [\hat{a}_\alpha^\dagger, \hat{N}] \hat{a}_\beta , \tag{2.2.1}$$

to break down the commutator of a product of creation and annihilation operators into commutators of individual creation and annihilation operators with $\hat{N}$. In the next step, we try to express the commutation relation with the annihilation operator in terms of the fundamental anticommutation relations for fermions,

$$[\hat{a}_\beta, \hat{N}] = \left[\hat{a}_\beta, \sum_\gamma \hat{a}_\gamma^\dagger \hat{a}_\gamma\right] = \sum_\gamma [\hat{a}_\beta, \hat{a}_\gamma^\dagger \hat{a}_\gamma]$$
$$= \sum_\gamma \left(\hat{a}_\beta \hat{a}_\gamma^\dagger \hat{a}_\gamma + \hat{a}_\gamma^\dagger \hat{a}_\beta \hat{a}_\gamma - \hat{a}_\gamma^\dagger \hat{a}_\beta \hat{a}_\gamma - \hat{a}_\gamma^\dagger \hat{a}_\gamma \hat{a}_\beta\right)$$
$$= \sum_\gamma \left(\{\hat{a}_\beta, \hat{a}_\gamma^\dagger\} \hat{a}_\gamma - \hat{a}_\gamma^\dagger \{\hat{a}_\beta, \hat{a}_\gamma\}\right) = \sum_\gamma \delta_{\beta\gamma} \hat{a}_\gamma = \hat{a}_\beta . \tag{2.2.2}$$

We follow similar steps for the commutator with the corresponding creation operator:

$$[\hat{a}_\alpha^\dagger, \hat{N}] = \left[\hat{a}_\alpha^\dagger, \sum_\gamma \hat{a}_\gamma^\dagger \hat{a}_\gamma\right] = \sum_\gamma [\hat{a}_\alpha^\dagger, \hat{a}_\gamma^\dagger \hat{a}_\gamma]$$
$$= \sum_\gamma \left(\hat{a}_\alpha^\dagger \hat{a}_\gamma^\dagger \hat{a}_\gamma + \hat{a}_\gamma^\dagger \hat{a}_\alpha^\dagger \hat{a}_\gamma - \hat{a}_\gamma^\dagger \hat{a}_\alpha^\dagger \hat{a}_\gamma - \hat{a}_\gamma^\dagger \hat{a}_\gamma \hat{a}_\alpha^\dagger\right)$$
$$= \sum_\gamma \{\hat{a}_\alpha^\dagger, \hat{a}_\gamma^\dagger\} \hat{a}_\gamma - \sum_\gamma \hat{a}_\gamma^\dagger \{\hat{a}_\alpha^\dagger, \hat{a}_\gamma\} = -\sum_\gamma \hat{a}_\gamma^\dagger \{\hat{a}_\gamma, \hat{a}_\alpha^\dagger\} = -\sum_\gamma \hat{a}_\gamma^\dagger \delta_{\alpha\gamma}$$
$$= -\hat{a}_\alpha^\dagger. \tag{2.2.3}$$

Using these commutators in Eq. (2.2.1), we find

$$[\hat{a}_\alpha^\dagger \hat{a}_\beta, \hat{N}] = \hat{a}_\alpha^\dagger [\hat{a}_\beta, \hat{N}] + [\hat{a}_\alpha^\dagger, \hat{N}] \hat{a}_\beta = \hat{a}_\alpha^\dagger \hat{a}_\beta - \hat{a}_\alpha^\dagger \hat{a}_\beta = 0.$$

The particle number operator thus commutes with any one-body Hamiltonian.

(b) For the commutator with the two-body part of the Hamiltonian, we proceed in a similar manner. First, we identify the terms within the relevant sums, $[\hat{H}_1, \hat{N}] = \frac{1}{2} \sum_{\alpha\beta\delta\gamma} v_{\alpha\beta\gamma\delta} [\hat{a}_\alpha^\dagger \hat{a}_\beta^\dagger \hat{a}_\delta \hat{a}_\gamma, \hat{N}]$. To evaluate this commutator, we add and remove terms to express it as commutators of individual creation and annihilation operators with $\hat{N}$:

$$
\begin{aligned}
\left[\hat{a}_\alpha^\dagger \hat{a}_\beta^\dagger \hat{a}_\delta \hat{a}_\gamma, \hat{N}\right] &= \hat{a}_\alpha^\dagger \hat{a}_\beta^\dagger \hat{a}_\delta \hat{a}_\gamma \hat{N} - \hat{N}\hat{a}_\alpha^\dagger \hat{a}_\beta^\dagger \hat{a}_\delta \hat{a}_\gamma \\
&= \hat{a}_\alpha^\dagger \hat{a}_\beta^\dagger \hat{a}_\delta \hat{a}_\gamma \hat{N} - \hat{a}_\alpha^\dagger \hat{a}_\beta^\dagger \hat{a}_\delta \hat{N} \hat{a}_\gamma + \hat{a}_\alpha^\dagger \hat{a}_\beta^\dagger \hat{a}_\delta \hat{N} \hat{a}_\gamma - \hat{N}\hat{a}_\alpha^\dagger \hat{a}_\beta^\dagger \hat{a}_\delta \hat{a}_\gamma \\
&= \hat{a}_\alpha^\dagger \hat{a}_\beta^\dagger \hat{a}_\delta \left[\hat{a}_\gamma, \hat{N}\right] + \hat{a}_\alpha^\dagger \hat{a}_\beta^\dagger \hat{a}_\delta \hat{N} \hat{a}_\gamma - \hat{a}_\alpha^\dagger \hat{a}_\beta^\dagger \hat{N} \hat{a}_\delta \hat{a}_\gamma + \hat{a}_\alpha^\dagger \hat{a}_\beta^\dagger \hat{N} \hat{a}_\delta \hat{a}_\gamma \\
&\quad - \hat{N}\hat{a}_\alpha^\dagger \hat{a}_\beta^\dagger \hat{a}_\delta \hat{a}_\gamma \\
&= \hat{a}_\alpha^\dagger \hat{a}_\beta^\dagger \hat{a}_\delta \left[\hat{a}_\gamma, \hat{N}\right] + \hat{a}_\alpha^\dagger \hat{a}_\beta^\dagger \left[\hat{a}_\delta, \hat{N}\right] \hat{a}_\gamma + \hat{a}_\alpha^\dagger \hat{a}_\beta^\dagger \hat{N} \hat{a}_\delta \hat{a}_\gamma \\
&\quad - \hat{a}_\alpha^\dagger \hat{N} \hat{a}_\beta^\dagger \hat{a}_\delta \hat{a}_\gamma + \hat{a}_\alpha^\dagger \hat{N} \hat{a}_\beta^\dagger \hat{a}_\delta \hat{a}_\gamma - \hat{N}\hat{a}_\alpha^\dagger \hat{a}_\beta^\dagger \hat{a}_\delta \hat{a}_\gamma \\
&= \hat{a}_\alpha^\dagger \hat{a}_\beta^\dagger \hat{a}_\delta \left[\hat{a}_\gamma, \hat{N}\right] + \hat{a}_\alpha^\dagger \hat{a}_\beta^\dagger \left[\hat{a}_\delta, \hat{N}\right] \hat{a}_\gamma + \hat{a}_\alpha^\dagger \left[\hat{a}_\beta^\dagger, \hat{N}\right] \hat{a}_\delta \hat{a}_\gamma \\
&\quad + \left[\hat{a}_\alpha^\dagger, \hat{N}\right] \hat{a}_\beta^\dagger \hat{a}_\delta \hat{a}_\gamma \\
&= \hat{a}_\alpha^\dagger \hat{a}_\beta^\dagger \hat{a}_\delta \hat{a}_\gamma + \hat{a}_\alpha^\dagger \hat{a}_\beta^\dagger \hat{a}_\delta \hat{a}_\gamma - \hat{a}_\alpha^\dagger \hat{a}_\beta^\dagger \hat{a}_\delta \hat{a}_\gamma - \hat{a}_\alpha^\dagger \hat{a}_\beta^\dagger \hat{a}_\delta \hat{a}_\gamma = 0.
\end{aligned}
$$

Taking into account that the commutator is a linear operation, we can thus write

$$
[\hat{H}, \hat{N}] = [\hat{H}_0, \hat{N}] + [\hat{H}_1, \hat{N}] = 0. \tag{2.2.4}
$$

This indicates that the two operators commute and hence can be diagonalized separately. This has two significant consequences. First, $\hat{N}$ is a constant of motion. Second, a common basis of many-body states can be built from the eigenstates of both $\hat{N}$ and $\hat{H}$. These eigenstates can be labeled by the eigenvalues of both $\hat{N}$ and $\hat{H}$ providing eigenenergies E_α^N.

(c) We can repeat the same steps followed in the fermionic case, but this time we want to express our commutators in terms of fundamental commutators (and not anticommutators) of annihilation and creation operators. This essentially reduces to recomputing the commutators in Eqs. (2.2.2) and (2.2.3):

$$
\begin{aligned}
[\hat{a}_\beta, \hat{N}] &= \left[\hat{a}_\beta, \sum_\gamma \hat{a}_\gamma^\dagger \hat{a}_\gamma\right] = \sum_\gamma [\hat{a}_\beta, \hat{a}_\gamma^\dagger \hat{a}_\gamma] = \sum_\gamma \left([\hat{a}_\beta, \hat{a}_\gamma^\dagger] \hat{a}_\gamma + \hat{a}_\gamma^\dagger [\hat{a}_\beta, \hat{a}_\gamma]\right) \\
&= \sum_\gamma \delta_{\beta\gamma} \hat{a}_\gamma = \hat{a}_\beta.
\end{aligned}
$$

This coincides with the results for fermions. Now, the commutator with the creation operator:

$$
\begin{aligned}
[\hat{a}_\alpha^\dagger, \hat{N}] &= \left[\hat{a}_\alpha^\dagger, \sum_\gamma \hat{a}_\gamma^\dagger \hat{a}_\gamma\right] = \sum_\gamma [\hat{a}_\alpha^\dagger, \hat{a}_\gamma^\dagger \hat{a}_\gamma] = \sum_\gamma \left([\hat{a}_\alpha^\dagger, \hat{a}_\alpha^\dagger] \hat{a}_\gamma + \hat{a}_\gamma^\dagger [\hat{a}_\alpha^\dagger, \hat{a}_\gamma]\right) \\
&= -\sum_\gamma \hat{a}_\gamma^\dagger \delta_{\alpha\gamma} = -\hat{a}_\alpha^\dagger,
\end{aligned}
$$

which also matches what was obtained for fermions. Now, given that the commutators with the number operator $\hat{N}$ for bosons are identical to their fermionic counterparts, it is immediate from the previous proof that any bosonic Hamiltonian $\hat{H}$ will also commute with the number of particles operator, $[\hat{H}, \hat{N}] = 0$.

PROBLEM 2.3
Anticommutation relations of field operators

We define the field operators as

$$\hat{\Psi}(\vec{r}) \equiv \sum_{\vec{k}} \varphi_{\vec{k}}(\vec{r}) \hat{a}_{\vec{k}} \quad , \quad \hat{\Psi}^\dagger(\vec{r}) \equiv \sum_{\vec{k}} \varphi_{\vec{k}}^*(\vec{r}) \hat{a}_{\vec{k}}^\dagger \quad , \tag{2.3.1}$$

where

$$\varphi_{\vec{k}}(\vec{r}) = \frac{1}{\Omega^{1/2}} e^{i\vec{k}\cdot\vec{r}} \tag{2.3.2}$$

are single-particle momentum eigenfunctions, normalized to a volume Ω assuming periodic boundary conditions. $\hat{a}_{\vec{k}}^\dagger$ $(\hat{a}_{\vec{k}})$ are the creation (annihilation) operators of single-particle states with a well-defined momentum $\vec{k}$. Prove that the field operators satisfy the following anticommutation relations:

(a) $\{\hat{\Psi}(\vec{r}), \hat{\Psi}(\vec{r}')\} = \{\hat{\Psi}^\dagger(\vec{r}), \hat{\Psi}^\dagger(\vec{r}')\} = 0$;

(b) $\{\hat{\Psi}(\vec{r}), \hat{\Psi}^\dagger(\vec{r}')\} = \delta(\vec{r} - \vec{r}')$.

(a) To find the corresponding operators, we use the definition of the field operators in terms of creation and annihilation operators:

$$\{\hat{\Psi}(\vec{r}), \hat{\Psi}(\vec{r}')\} = \left\{ \sum_{\vec{k}} \varphi_{\vec{k}}(\vec{r}) \hat{a}_{\vec{k}}, \sum_{\vec{k}'} \varphi_{\vec{k}'}(\vec{r}') \hat{a}_{\vec{k}'} \right\} = \sum_{\vec{k}} \sum_{\vec{k}'} \varphi_{\vec{k}}(\vec{r}) \varphi_{\vec{k}'}(\vec{r}') \{\hat{a}_{\vec{k}}, \hat{a}_{\vec{k}'}\} .$$

Then, using the anticommutation relations of the creation and annihilation operators, $\{\hat{a}_{\vec{k}}, \hat{a}_{\vec{k}'}\} = 0$, we conclude that

$$\{\hat{\Psi}(\vec{r}), \hat{\Psi}(\vec{r}')\} = 0 .$$

Taking Hermitian conjugates, or working with a very similar argument, leads to the same result for the conjugate operators, $\{\hat{\Psi}^\dagger(\vec{r}), \hat{\Psi}^\dagger(\vec{r}')\}$.

(b) To find the mixed anticommutator, we use similar mathematical machinery, starting from

$$\{\Psi(\vec{r}), \hat{\Psi}^\dagger(\vec{r}')\} = \left\{ \sum_{\vec{k}} \varphi_{\vec{k}}(\vec{r}) \hat{a}_{\vec{k}}, \sum_{\vec{k}'} \varphi_{\vec{k}'}^*(\vec{r}) \hat{a}_{\vec{k}}^\dagger \right\} = \sum_{\vec{k}} \sum_{\vec{k}'} \varphi_{\vec{k}}(\vec{r}) \varphi_{\vec{k}'}^*(\vec{r}') \{\hat{a}_{\vec{k}}, \hat{a}_{\vec{k}'}^\dagger\} .$$

Then, taking into account that $\left\{\hat{a}_{\vec{k}}, \hat{a}_{\vec{k}'}^{\dagger}\right\} = \delta_{\vec{k}\vec{k}'}$ one gets that

$$\left\{\Psi(\vec{r}), \hat{\Psi}^{\dagger}(\vec{r}')\right\} = \sum_{\vec{k}}\sum_{\vec{k}'} \varphi_{\vec{k}}(\vec{r})\varphi_{\vec{k}'}^{*}(\vec{r}')\delta_{\vec{k}\vec{k}'} = \sum_{\vec{k}} \varphi_{\vec{k}}(\vec{r})\varphi_{\vec{k}}^{*}(\vec{r}') = \delta(\vec{r}-\vec{r}'),$$

due to the completeness relation of the wavefunctions $\varphi_{\vec{k}}(\vec{r})$.

PROBLEM 2.4
Particle density operator

The density operator for a system of N particles in first quantization is expressed as

$$\rho(\vec{r}) = \sum_{i=1}^{N} \delta(\vec{r}-\vec{r}_i). \qquad (2.4.1)$$

(a) Show that the same operator in second quantization is $\hat{\rho}(\vec{r}) = \hat{\Psi}^{\dagger}(\vec{r})\hat{\Psi}(\vec{r})$, where $\hat{\Psi}(\vec{r}) = \sum_l \varphi_l(\vec{r})\hat{a}_l$ is the field operator.

(b) Show that the particle number operator $\hat{N} = \sum_l \hat{a}_l^{\dagger}\hat{a}_l$ can be expressed as $\hat{N} = \int d\vec{r}\hat{\rho}(\vec{r})$. Note that $\hat{a}_l^{\dagger}(\hat{a}_l)$ are the creation (annihilation) operators corresponding to a complete single-particle basis.

(a) In second quantization, the expression for any one-body operator is given, see Eq. (A.0.12), by

$$\hat{\rho}(\vec{r}) = \sum_{lm} \langle l|\delta(\vec{r}-\vec{r}_i)|m\rangle \hat{a}_l^{\dagger}\hat{a}_m,$$

where $\langle l|\delta(\vec{r}-\vec{r}_i)|m\rangle$ are the one-body matrix elements of the single-particle component of the operator, which is the same independently of the value of i. We employ a complete single-particle basis $\varphi_l(\vec{r})$, with associated creation (destruction) operators $\hat{a}_l^{\dagger}(\hat{a}_l)$. With this in mind, the previous expression can be written as

$$\hat{\rho}(\vec{r}) = \sum_{lm} \hat{a}_l^{\dagger}\hat{a}_m \int d\vec{r}_i \varphi_l^{*}(\vec{r}_i)\delta(\vec{r}-\vec{r}_i)\varphi_m(\vec{r}_i) = \sum_{lm} \hat{a}_l^{\dagger}\hat{a}_m \varphi_l^{*}(\vec{r})\varphi_m(\vec{r})$$

$$= \left(\sum_l \varphi_l^{*}(\vec{r})\hat{a}_l^{\dagger}\right)\left(\sum_m \varphi_m(\vec{r})\hat{a}_m\right) = \hat{\Psi}^{\dagger}(\vec{r})\hat{\Psi}(\vec{r}). \qquad (2.4.2)$$

It is important to note that the integration is performed over the single-particle coordinate $\vec{r}_i$, whereas the position $\vec{r}$ represents the point at which the density is evaluated.

(b) We start from the integral and use the result of the previous section to find:

$$\int d\vec{r}\,\hat{\rho}(\vec{r}) = \int d\vec{r}\,\hat{\Psi}^\dagger(\vec{r})\hat{\Psi}(\vec{r}) = \int d\vec{r}\left(\sum_l \varphi_l^*(\vec{r})\hat{a}_l^\dagger\right)\left(\sum_m \varphi_m(\vec{r})\hat{a}_m\right)$$

$$= \sum_{lm}\hat{a}_l^\dagger \hat{a}_m \int d\vec{r}\,\varphi_l^*(\vec{r})\varphi_m(\vec{r}) = \sum_{lm}\hat{a}_l^\dagger \hat{a}_m \delta_{lm} = \sum_l \hat{a}_l^\dagger \hat{a}_l \equiv \hat{N}, \quad (2.4.3)$$

where we have used the normalization of the single-particle wavefunctions.

PROBLEM 2.5
The states generated by field operators are eigenvectors of the particle number operator

Using the anticommutation relations of the field operators $\hat{\Psi}^\dagger(\vec{r})$ and $\hat{\Psi}(\vec{r})$ of a system of identical fermions, show that

(a) $\hat{\Psi}^\dagger(\vec{r})|0\rangle$ is an eigenstate of the particle number operator.

(b) $\hat{\Psi}^\dagger(\vec{r})\hat{\Psi}^\dagger(\vec{r}')|0\rangle$ is also an eigenstate of $\hat{N}$.

In both cases, determine the corresponding eigenvalues.

(a) We can directly apply $\hat{N}$ on $\hat{\Psi}^\dagger(\vec{r}')|0\rangle$ and manipulate the expression to find:

$$\left[\int d\vec{r}\,\hat{\Psi}^\dagger(\vec{r})\hat{\Psi}(\vec{r})\right]\hat{\Psi}^\dagger(\vec{r}')|0\rangle = \int d\vec{r}\,\hat{\Psi}^\dagger(\vec{r})\left(\delta(\vec{r}-\vec{r}') - \hat{\Psi}^\dagger(\vec{r}')\hat{\Psi}(\vec{r})\right)|0\rangle$$

$$= \hat{\Psi}^\dagger(\vec{r}')|0\rangle. \quad (2.5.1)$$

Therefore, $\hat{\Psi}^\dagger(\vec{r})|0\rangle$ is an eigenvector of $\hat{N}$ with eigenvalue 1.

(b) Again, we can directly apply $\hat{N}$ onto the state and see that:

$$\left[\int d\vec{r}\,\hat{\Psi}^\dagger(\vec{r})\hat{\Psi}(\vec{r})\right]\hat{\Psi}^\dagger(\vec{r}')\hat{\Psi}^\dagger(\vec{r}'')|0\rangle$$

$$= \int d\vec{r}\,\hat{\Psi}^\dagger(\vec{r})\left(-\hat{\Psi}^\dagger(\vec{r}')\hat{\Psi}(\vec{r}) + \delta(\vec{r}-\vec{r}')\right)\hat{\Psi}^\dagger(\vec{r}'')|0\rangle$$

$$= \int d\vec{r}\,\hat{\Psi}^\dagger(\vec{r})\delta(\vec{r}-\vec{r}')\hat{\Psi}^\dagger(\vec{r}'')|0\rangle - \int d\vec{r}\,\hat{\Psi}^\dagger(\vec{r})\hat{\Psi}^\dagger(\vec{r}')\hat{\Psi}(\vec{r})\hat{\Psi}^\dagger(\vec{r}'')|0\rangle$$

$$= \hat{\Psi}^\dagger(\vec{r}')\hat{\Psi}^\dagger(\vec{r}'')|0\rangle - \int d\vec{r}\,\hat{\Psi}^\dagger(\vec{r})\hat{\Psi}^\dagger(\vec{r}')\left(\delta(\vec{r}-\vec{r}'') - \hat{\Psi}^\dagger(\vec{r}'')\hat{\Psi}(\vec{r})\right)|0\rangle$$

$$= \hat{\Psi}^\dagger(\vec{r}')\hat{\Psi}^\dagger(\vec{r}'')|0\rangle - \hat{\Psi}^\dagger(\vec{r}'')\hat{\Psi}^\dagger(\vec{r}')|0\rangle$$

$$= \hat{\Psi}^\dagger(\vec{r}')\hat{\Psi}^\dagger(\vec{r}'')|0\rangle + \hat{\Psi}^\dagger(\vec{r}')\hat{\Psi}^\dagger(\vec{r}'')|0\rangle$$

$$= 2\hat{\Psi}^\dagger(\vec{r}')\hat{\Psi}^\dagger(\vec{r}'')|0\rangle. \quad (2.5.2)$$

Therefore, $\hat{\Psi}^{\dagger}(\vec{r}')\hat{\Psi}^{\dagger}(\vec{r}'')|0\rangle$ is an eigenvector of $\hat{N}$ with eigenvalue 2.

PROBLEM 2.6
Heisenberg representation of creation and destruction operators for a single-particle Hamiltonian

Consider a Hamiltonian which is diagonal in a given single-particle basis:

$$\hat{H}_0 = \sum_{\alpha} \varepsilon_{\alpha} \hat{a}^{\dagger}_{\alpha} \hat{a}_{\alpha}. \qquad (2.6.1)$$

Find $\hat{a}_{\alpha}$ and $\hat{a}^{\dagger}_{\alpha}$ in the Heisenberg representation.

We start with the definition of a Heisenberg picture annihilation operator,

$$\hat{a}_{\alpha,H}(t) = e^{i\frac{\hat{H}_0 t}{\hbar}} \, \hat{a}_{\alpha} \, e^{-i\frac{\hat{H}_0 t}{\hbar}}.$$

This definition can be expanded in powers of t in terms of nested commutators,

$$\hat{a}_{\alpha,H}(t) = \hat{a}_{\alpha} + i[\hat{H}_0,\hat{a}_{\alpha}]\frac{t}{\hbar} + \frac{i^2}{2!}[\hat{H}_0,[\hat{H}_0,\hat{a}_{\alpha}]]\frac{t^2}{\hbar^2} + \dots. \qquad (2.6.2)$$

We evaluate the first commutator,

$$
\begin{aligned}
\left[\hat{H}_0,\hat{a}_{\alpha}\right] &= \left[\sum_{\gamma} \varepsilon_{\gamma}\hat{a}^{\dagger}_{\gamma}\hat{a}_{\gamma},\hat{a}_{\alpha}\right] = \sum_{\gamma}\varepsilon_{\gamma}[\hat{a}^{\dagger}_{\gamma}\hat{a}_{\gamma},\hat{a}_{\alpha}] = \sum_{\gamma}\varepsilon_{\gamma}(\hat{a}^{\dagger}_{\gamma}\hat{a}_{\gamma}\hat{a}_{\alpha} - \hat{a}_{\alpha}\hat{a}^{\dagger}_{\gamma}\hat{a}_{\gamma}) \\
&= \sum_{\gamma}\varepsilon_{\gamma}(\hat{a}^{\dagger}_{\gamma}\hat{a}_{\gamma}\hat{a}_{\alpha} + \hat{a}^{\dagger}_{\gamma}\hat{a}_{\alpha}\hat{a}_{\gamma} - \hat{a}^{\dagger}_{\gamma}\hat{a}_{\alpha}\hat{a}_{\gamma} - \hat{a}_{\alpha}\hat{a}^{\dagger}_{\gamma}\hat{a}_{\gamma}) \\
&= \sum_{\gamma}\varepsilon_{\gamma}\left(\hat{a}^{\dagger}_{\gamma}\{\hat{a}_{\gamma},\hat{a}_{\alpha}\} - \{\hat{a}^{\dagger}_{\gamma},\hat{a}_{\alpha}\}\hat{a}_{\gamma}\right) = -\sum_{\gamma}\varepsilon_{\gamma}\delta_{\gamma\alpha}\hat{a}_{\alpha} \\
&= -\varepsilon_{\alpha}\hat{a}_{\alpha},
\end{aligned}
$$

employing the fermionic anticommutation relations. In other words, the first commutator is just

$$[\hat{H}_0,\hat{a}_{\alpha}] = -\varepsilon_{\alpha}\hat{a}_{\alpha}.$$

The second commutator in Eq. (2.6.2) can easily be computed from this result,

$$[\hat{H}_0,[\hat{H}_0,\hat{a}_{\alpha}]] = [\hat{H}_0,-\varepsilon_{\alpha}\hat{a}_{\alpha}] = -\varepsilon_{\alpha}[\hat{H}_0,\hat{a}_{\alpha}] = \varepsilon^2_{\alpha}\hat{a}_{\alpha}.$$

In turn, the $n-$th order nested commutator is simply

$$\underbrace{[\hat{H}_0,[\hat{H}_0,\dots,[\hat{H}_0,\hat{a}_{\alpha}]]\dots]}_{n \text{ times}} = (-1)^n\varepsilon^n_{\alpha}\hat{a}_{\alpha}.$$

Taking into account this result, one can write the complete series expansion on t,

$$\hat{a}_{\alpha,H}(t) = \hat{a}_\alpha - i\varepsilon_\alpha \frac{t}{\hbar}\hat{a}_\alpha + \frac{i^2}{2!}\frac{t^2}{\hbar^2}\varepsilon_\alpha^2\hat{a}_\alpha - \frac{i^3}{3!}\varepsilon_\alpha^3\frac{t^3}{\hbar^3}\hat{a}_\alpha + \cdots$$

$$= \hat{a}_\alpha\left(1 - i\varepsilon_\alpha\frac{t}{\hbar} + \frac{i^2}{2!}\varepsilon_\alpha^2 - \cdots\right), \qquad (2.6.3)$$

and therefore,

$$\hat{a}_{\alpha,H}(t) = \hat{a}_\alpha\, e^{-i\frac{\varepsilon_\alpha t}{\hbar}}. \qquad (2.6.4)$$

This gives the Heisenberg picture version of the destruction operator.

We could have obtained this result with an alternative procedure. One may write the equation of motion of the destruction operator $\hat{a}_{\alpha,H}$:

$$i\hbar\frac{d}{dt}\hat{a}_{\alpha,H}(t) = [\hat{a}_{\alpha,H}(t), \hat{H}_0] = \left[e^{i\hat{H}_0 t/\hbar}\hat{a}_\alpha e^{-i\hat{H}_0 t/\hbar}, \hat{H}_0\right]$$

$$= e^{i\hat{H}_0 t/\hbar}\left[\hat{a}_\alpha, \sum_\gamma \varepsilon_\gamma \hat{a}_\gamma^\dagger \hat{a}_\gamma\right] e^{-i\hat{H}_0 t/\hbar}$$

$$= e^{i\hat{H}_0 t/\hbar} \sum_\gamma \varepsilon_\gamma \left[\hat{a}_\alpha, \hat{a}_\gamma^\dagger \hat{a}_\gamma\right] e^{-i\hat{H}_0 t/\hbar}.$$

Now we calculate the commutator

$$\left[\hat{a}_\alpha, \hat{a}_k^\dagger \hat{a}_k\right] = \delta_{k\alpha}\hat{a}_\alpha,$$

and, substituting in the previous expression, we have:

$$i\hbar\frac{d}{dt}\hat{a}_{\alpha,H}(t) = e^{i\hat{H}_0 t/\hbar} \sum_k \varepsilon_k \delta_{k\alpha}\hat{a}_\alpha\, e^{-i\hat{H}_0 t/\hbar} = e^{i\hat{H}_0 t/\hbar}\, \varepsilon_\alpha\, \hat{a}_\alpha\, e^{-i\hat{H}_0 t/\hbar} = \varepsilon_\alpha\, \hat{a}_{\alpha,H}(t).$$

This indicates that the destruction operator in the Heisenberg picture fulfills the following closed first-order differential equation:

$$i\hbar\frac{d}{dt}\hat{a}_{\alpha,H}(t) = \varepsilon_\alpha\, \hat{a}_{\alpha,H}(t). \qquad (2.6.5)$$

The solution of this equation is straightforward:

$$\hat{a}_{\alpha,H}(t) = \hat{a}_\alpha\, e^{-i\varepsilon_\alpha t/\hbar}, \qquad (2.6.6)$$

with the boundary condition for $t = 0$, $\hat{a}_{\alpha,H}(t=0) = \hat{a}_\alpha$. This is the same result obtained in Eq. (2.6.4).

We could follow either of these two methods to find the equivalent expression for the creation operator $\hat{a}_{\alpha,H}^\dagger(t)$. This, however, can be obtained directly from the Hermitian conjugate of Eq. (2.6.4), so we get

$$\hat{a}_{\alpha,H}^\dagger(t) = \hat{a}_\alpha^\dagger\, e^{i\varepsilon_\alpha t/\hbar}. \qquad (2.6.7)$$

PROBLEM 2.7
A canonical linear transformation

Show that a linear transformation of the creation and destruction operators, defined by

$$\hat{d}_k^\dagger = \sum_l D_{lk} \hat{a}_l^\dagger \quad , \quad \hat{d}_{k'} = \sum_l D_{lk'}^* \hat{a}_l, \tag{2.7.1}$$

is a canonical transformation , i.e. it preserves the anticommutation relations of the creation and annihilation operators, as long as D is unitary.

We first compute the anticommutation relations of the new set of creation operators,

$$\left\{ \hat{d}_k^\dagger, \hat{d}_{k'}^\dagger \right\} = \left\{ \sum_n D_{nk} \, \hat{a}_n^\dagger, \sum_m D_{mk}^* \, \hat{a}_m^\dagger \right\} = \sum_n \sum_m D_{nk} D_{mk'} \left\{ \hat{a}_n^\dagger, \hat{a}_m^\dagger \right\} = 0. \tag{2.7.2}$$

This is the expected result. We can do a similar calculation for the destruction operators, to find $\left\{ \hat{d}_k, \hat{d}_{k'} \right\} = 0$. Note that these anticommutation relations are preserved for any linear transformation, independently of whether it is unitary or not.

We now compute the mixed anticommutator, and see that

$$\left\{ \hat{d}_k, \hat{d}_{k'}^\dagger \right\} = \left\{ \sum_n D_{nk}^* \, \hat{a}_n, \sum_m D_{mk'} \, \hat{a}_m^\dagger \right\} = \sum_{nm} D_{nk}^* D_{mk'} \left\{ \hat{a}_n, \hat{a}_m^\dagger \right\}$$

$$= \sum_{nm} D_{nk}^* D_{mk'} \delta_{nm} = \sum_n D_{nk}^* D_{nk'} = \sum_n (D^\dagger)_{kn} (D)_{nk'}$$

$$= (D^\dagger D)_{kk'} = \delta_{kk'} . \tag{2.7.3}$$

We have exploited the fact that D is unitary in the last line.

PROBLEM 2.8
Spin operators in second quantization

Consider a uniform system of fermions of spin $S = 1/2$. We build a Fock space using a single-particle basis of plane waves, i.e. states with good linear momentum $\vec{k}$ and spin projections $(\uparrow, \downarrow)$ on the z-axis. We associate to these single-particle states the corresponding creation (annihilation) operators $\hat{a}_{\vec{k}\uparrow}^\dagger, \hat{a}_{\vec{k}\downarrow}^\dagger, \hat{a}_{\vec{k}\uparrow}$, and $\hat{a}_{\vec{k}\downarrow}$.

(a) Prove that the components of the total spin operator, $\vec{S}$, are given by:

$$\hat{S}_x = \frac{1}{2} \sum_{\vec{k}} \left[\hat{a}_{\vec{k}\uparrow}^{\dagger} \hat{a}_{\vec{k}\downarrow} + \hat{a}_{\vec{k}\downarrow}^{\dagger} \hat{a}_{\vec{k}\uparrow} \right],$$

$$\hat{S}_y = \frac{-i}{2} \sum_{\vec{k}} \left[\hat{a}_{\vec{k}\uparrow}^{\dagger} \hat{a}_{\vec{k}\downarrow} - \hat{a}_{\vec{k}\downarrow}^{\dagger} \hat{a}_{\vec{k}\uparrow} \right],$$

$$\hat{S}_z = \frac{1}{2} \sum_{\vec{k}} \left[\hat{a}_{\vec{k}\uparrow}^{\dagger} \hat{a}_{\vec{k}\uparrow} - \hat{a}_{\vec{k}\downarrow}^{\dagger} \hat{a}_{\vec{k}\downarrow} \right]. \qquad (2.8.1)$$

(b) Calculate the expectation value of $\hat{S}_z, \hat{S}_x, \hat{S}_y$ on the unpolarized Fermi sea,

$$|\Psi_{FS}\rangle = \prod_{k \leq k_F} \hat{a}_{\vec{k}\uparrow}^{\dagger} \prod_{k \leq k_F} \hat{a}_{\vec{k}\downarrow}^{\dagger} |0\rangle. \qquad (2.8.2)$$

(c) Calculate the expectation value of $\hat{S}_z, \hat{S}_x, \hat{S}_y$ on the polarized Fermi sea,

$$|\Psi_{PFS}\rangle = \prod_{k \leq k_F} \hat{a}_{\vec{k}\uparrow}^{\dagger} |0\rangle. \qquad (2.8.3)$$

(a) The operators $\hat{S}_x, \hat{S}_y$, and $\hat{S}_z$ are one-body operators. We shall employ Eq. (A.0.12) to transform each operator into the second quantization representation. We start with $\hat{S}_z$, which in second quantization reads

$$\hat{S}_z = \sum_{\vec{k}\sigma, \vec{l}\sigma'} \langle \vec{k}\sigma | \hat{s}_z | \vec{l}\sigma' \rangle \, \hat{a}_{\vec{k}\sigma}^{\dagger} \hat{a}_{\vec{l}\sigma'}. \qquad (2.8.4)$$

Here, $\hat{s}_z = (1/2)\hat{\sigma}_z$, with $\hat{\sigma}_z$ the corresponding Pauli matrix. With this, the one-body matrix elements read

$$\langle \vec{k}\sigma | \hat{s}^z | \vec{l}\sigma' \rangle = \langle \vec{k} | \vec{l} \rangle \langle \sigma | \hat{s}^z | \sigma' \rangle = \frac{1}{2} \delta_{\vec{k}\vec{l}} \, (\hat{\sigma}^z)_{\sigma\sigma'}, \qquad (2.8.5)$$

where $(\hat{\sigma}^k)_{\sigma\sigma'}$ denote the matrix elements of the k-th Pauli matrix, such that $(\hat{\sigma}^k)_{\uparrow\uparrow} = +1$ and $(\hat{\sigma}^k)_{\downarrow\downarrow} = -1$. With this, we find that the operator is expressed as

$$\hat{S}_z = \frac{1}{2} \sum_{\vec{k}} \left[\hat{a}_{\vec{k}\uparrow}^{\dagger} \hat{a}_{\vec{k}\uparrow} - \hat{a}_{\vec{k}\downarrow}^{\dagger} \hat{a}_{\vec{k}\downarrow} \right]. \qquad (2.8.6)$$

We now proceed to find the second quantization expression for $\hat{S}_x$. We start once again from Eq. (A.0.12),

$$\hat{S}_x = \sum_{\vec{k}\sigma, \vec{l}\sigma'} \langle \vec{k}\sigma | \hat{s}^x | \vec{l}\sigma' \rangle \, \hat{a}_{\vec{k}\sigma}^{\dagger} \hat{a}_{\vec{l}\sigma'}, \qquad (2.8.7)$$

and the one-body matrix elements are now related to the Pauli matrix $\hat{\sigma}_x$,

$$\langle \vec{k}\sigma|\hat{s}^x|\vec{l}\sigma'\rangle = \langle \vec{k}|\vec{l}\rangle \langle \sigma|\hat{s}^x|\sigma'\rangle = \frac{1}{2}\delta_{\vec{k}\vec{l}}\,(\hat{\sigma}^x)_{\sigma\sigma'}. \tag{2.8.8}$$

Using the fact that

$$(\hat{\sigma}^x)_{\sigma\sigma'} = \delta_{\sigma-\sigma'}, \tag{2.8.9}$$

we find that

$$\hat{S}_x = \frac{1}{2}\sum_{\vec{k}}\left[\hat{a}^\dagger_{\vec{k}\uparrow}\hat{a}_{\vec{k}\downarrow} + \hat{a}^\dagger_{\vec{k}\downarrow}\hat{a}_{\vec{k}\uparrow}\right]. \tag{2.8.10}$$

Finally, for $\hat{S}_y$, the expression reads

$$\hat{S}_y = \sum_{\vec{k}\sigma,\vec{l}\sigma'}\langle \vec{k}\sigma|\hat{s}^y|\vec{l}\sigma'\rangle\,\hat{a}^\dagger_{\vec{k}\sigma}\hat{a}_{\vec{l}\sigma'}. \tag{2.8.11}$$

In this case, the one-body matrix elements are related to the $\hat{\sigma}^y$ Pauli matrix,

$$\langle \vec{k}\sigma|\hat{s}^y|\vec{l}\sigma'\rangle = \langle \vec{k}|\vec{l}\rangle \langle \sigma|\hat{s}^y|\sigma'\rangle = \frac{1}{2}\delta_{\vec{k}\vec{l}}\,(\hat{\sigma}^y)_{\sigma\sigma'}. \tag{2.8.12}$$

Now, for this matrix we know that $\hat{\sigma}^y_{\downarrow,\uparrow} = -i$ and $\hat{\sigma}^y_{\uparrow,\downarrow} = i$, which translates into the following second quantization expression

$$\hat{S}_y = \frac{-i}{2}\sum_{\vec{k}}\left[\hat{a}^\dagger_{\vec{k}\uparrow}\hat{a}_{\vec{k}\downarrow} - \hat{a}^\dagger_{\vec{k}\downarrow}\hat{a}_{\vec{k}\uparrow}\right]. \tag{2.8.13}$$

(b) We now proceed to find the expectation value of each total spin component in an unpolarized free Fermi sea. We start with the $\hat{S}_z$ component:

$$\langle \Psi_{FS}|\hat{S}_z|\Psi_{FS}\rangle = \left\langle \Psi_{FS}\left|\frac{1}{2}\sum_{\vec{k}}\left[\hat{a}^\dagger_{\vec{k}\uparrow}\hat{a}_{\vec{k}\uparrow} - \hat{a}^\dagger_{\vec{k}\downarrow}\hat{a}_{\vec{k}\downarrow}\right]\right|\Psi_{FS}\right\rangle$$
$$= \frac{N}{2} - \frac{N}{2} = 0. \tag{2.8.14}$$

This result is a direct consequence of the fact that the operator $\hat{a}^\dagger_{\vec{k}\uparrow}\hat{a}_{\vec{k}\uparrow}$ counts the number of particles with spin up, which by construction in the unpolarized Fermi sea is $N/2$. Similarly, the operator $\hat{a}^\dagger_{\vec{k}\downarrow}\hat{a}_{\vec{k}\downarrow}$ counts the number of particles with spin down, which by construction is also $N/2$. Overall, the expectation value is zero. In other words, $|\Psi_{FS}\rangle$ is an eigenstate of $\hat{S}_z$ with a zero eigenvalue.

Let us analyze now the expectation value of $\hat{S}_x$. We use the expression of Eq. (2.8.10) and compute the expectation value to find

$$\langle \Psi_{FS}|\hat{S}_x|\Psi_{FS}\rangle = \langle \Psi_{FS}|\frac{1}{2}\sum_{\vec{k}}\left[\hat{a}^\dagger_{\vec{k}\uparrow}\hat{a}_{\vec{k}\downarrow} + \hat{a}^\dagger_{\vec{k}\downarrow}\hat{a}_{\vec{k}\uparrow}\right]|\Psi_{FS}\rangle = 0. \tag{2.8.15}$$

This result arises because, in the first term, we destroy a particle with momentum $\vec{k}$ and spin projection $\downarrow$ below the Fermi sea. If we try to create a particle with the same momentum (below the Fermi sea) and opposite spin projection, the state is already occupied by construction of the free Fermi sea. The Pauli principle does not allow this operation. The same is true for the second operator, which destroys a particle with up spin and tries to create one with a down component.

This very same argument applies to the $\hat{S}_y$ operator. Indeed, Eq. (2.8.13) is just a different linear combination of the same operators involved in $\hat{S}_x$, Eq. (2.8.10). Consequently, the same arguments are valid and one easily proves that

$$\langle \Psi_{\text{FS}} | \hat{S}_y | \Psi_{\text{FS}} \rangle = 0. \qquad (2.8.16)$$

(c) We build the polarized Fermi, from a state built of spins pointing upwards, $|\Psi_{\text{PFS}}\rangle = \prod_{k \leq k_F} \hat{a}^\dagger_{\vec{k}\uparrow} |0\rangle$. The expectation value of $\hat{S}_z$ in this state thus reads

$$\langle \Psi_{\text{PFS}} | \hat{S}_z | \Psi_{\text{PFS}} \rangle = \langle \Psi_{\text{PFS}} | \frac{1}{2} \sum_{\vec{k}} \left[\hat{a}^\dagger_{\vec{k}\uparrow} \hat{a}_{\vec{k}\uparrow} - \hat{a}^\dagger_{\vec{k}\downarrow} \hat{a}_{\vec{k}\downarrow} \right] | \Psi_{\text{PFS}} \rangle = \frac{N}{2}. \qquad (2.8.17)$$

The operator counts only the particles with spin $\uparrow$, which are the ones present in the spin-polarized system.

However, for $\hat{S}_x$ and $\hat{S}_y$ one can apply similar arguments to that given in the previous section in terms of the mixed products of creation and destruction operators. With the combination $a^\dagger_{\vec{k}\downarrow} \hat{a}_{\vec{k}\uparrow}$, for instance, one destroys a particle with spin $\uparrow$ and creates one with spin $\downarrow$. The corresponding state is, however, orthogonal to the spin-up state $|\Psi_{\text{PFS}}\rangle$. The opposite term, $a^\dagger_{\vec{k}\uparrow} \hat{a}_{\vec{k}\downarrow}$, tries to destroy a particle with spin $\downarrow$, which does not exist in $|\Psi_{\text{PFS}}\rangle$. All in all, we find that both expectation values are zero. For instance,

$$\langle \Psi_{\text{PFS}} | \hat{S}_x | \Psi_{\text{PFS}} \rangle = \langle \Psi_{\text{PFS}} | \frac{1}{2} \sum_{\vec{k}} \left[\hat{a}^\dagger_{\vec{k}\uparrow} \hat{a}_{\vec{k}\downarrow} + \hat{a}^\dagger_{\vec{k}\downarrow} \hat{a}_{\vec{k}\uparrow} \right] | \Psi_{\text{PFS}} \rangle = 0. \qquad (2.8.18)$$

The same arguments apply for $\hat{S}_y$ and we indeed get:

$$\langle \Psi_{\text{PFS}} | \hat{S}_y | \Psi_{\text{PFS}} \rangle = 0. \qquad (2.8.19)$$

3 Wick's theorem

In the context of many-body quantum mechanics, one often encounters the need to evaluate expectation values of operators or their matrix elements. For instance, given a Hamiltonian $\hat{H}$, we typically want to evaluate its expectation value, which is the energy of the system, $E = \langle \hat{H} \rangle$. In second quantization, the matrix element of an operator can always be expressed as a vacuum expectation value. This generally yields an expression which is hard to evaluate. Fortunately, a powerful tool to simplify these calculations exists: Wick's theorem. This theorem provides a systematic way to break down expressions involving products of creation and annihilation operators into sums of simpler terms.

The statement of the theorem is as follows:

$$
\begin{aligned}
\hat{A}_1 \ldots \hat{A}_n =& \mathcal{N}(\hat{A}_1 \ldots \hat{A}_n) \\
&+ \sum_{(ij)} \mathcal{N}(\hat{A}_1 \ldots \overline{\hat{A}_i \ldots \hat{A}_j} \ldots \hat{A}_n) \\
&+ \sum_{(ij)(kl)} \mathcal{N}(\hat{A}_1 \ldots \overline{\hat{A}_i \ldots \hat{A}_k \ldots \hat{A}_j \ldots \hat{A}_l} \ldots \hat{A}_n) \\
&\ \ \vdots \\
&+ \text{sum over fully-contracted products}.
\end{aligned}
$$

Here, $\hat{A}_i$ denotes an arbitrary creation or destruction operator, and $\mathcal{N}$ is the normal-ordering operator. This ordering is such that all creation operators are to the left of all destruction operators. In addition, Wick contractions of two operators $\hat{A}_i, \hat{A}_j$ are indicated with $\overline{\hat{A}_i \hat{A}_j}$.

In essence, Wick's theorem separates the products of creation and destruction operators into their normal-ordered part, plus all the possible normal-ordered contractions (which may involve a phase in the fermionic case). When evaluated on the vacuum, normal-ordered strings of operators are typically zero, and only fully-contracted terms remain. We refer the reader to Appendix B for a more detailed explanation of Wick's theorem.

PROBLEM 3.1
Contraction of two creation and destruction operators

Consider a basis of single-particle states $(\alpha, \beta, \gamma, ...)$ associated to a fermionic system, where α destruction represents one or more quantum numbers necessary to characterize the single-particle states. Consider now the creation and destruction operators corresponding to this single-particle basis, that fulfill the standard anticommutation relations. For two generic operators $\hat{a}$ and $\hat{b}$, their contraction is defined as:

$$\overset{\frown}{\hat{a}\hat{b}} = \hat{a}\hat{b} - \mathcal{N}(\hat{a}\hat{b}) \tag{3.1.1}$$

(a) Calculate the following contractions and their expectation values on the vacuum:

$$\overset{\frown}{\hat{a}_\alpha\hat{a}_\beta}, \overset{\frown}{\hat{a}_\alpha^\dagger\hat{a}_\beta^\dagger}, \overset{\frown}{\hat{a}_\alpha^\dagger\hat{a}_\beta}, \overset{\frown}{\hat{a}_\alpha\hat{a}_\beta^\dagger}. \tag{3.1.2}$$

(b) What is the expectation value in the vacuum of a product of an odd number of creation and destruction operators?

(a) We compute the contraction of the four terms one by one:

$$\overset{\frown}{\hat{a}_\alpha\hat{a}_\beta} = \hat{a}_\alpha\hat{a}_\beta - \mathcal{N}(\hat{a}_\alpha\hat{a}_\beta) = \hat{a}_\alpha\hat{a}_\beta - \hat{a}_\alpha\hat{a}_\beta = 0, \tag{3.1.3}$$

$$\overset{\frown}{\hat{a}_\alpha^\dagger\hat{a}_\beta^\dagger} = \hat{a}_\alpha^\dagger\hat{a}_\beta^\dagger - \mathcal{N}(\hat{a}_\alpha^\dagger\hat{a}_\beta^\dagger) = \hat{a}_\alpha^\dagger\hat{a}_\beta^\dagger - \hat{a}_\alpha^\dagger\hat{a}_\beta^\dagger = 0, \tag{3.1.4}$$

$$\overset{\frown}{\hat{a}_\alpha^\dagger\hat{a}_\beta} = \hat{a}_\alpha^\dagger\hat{a}_\beta - \mathcal{N}(\hat{a}_\alpha^\dagger\hat{a}_\beta) = \hat{a}_\alpha^\dagger\hat{a}_\beta - \hat{a}_\alpha^\dagger\hat{a}_\beta = 0, \tag{3.1.5}$$

$$\overset{\frown}{\hat{a}_\alpha\hat{a}_\beta^\dagger} = \hat{a}_\alpha\hat{a}_\beta^\dagger - \mathcal{N}(\hat{a}_\alpha\hat{a}_\beta^\dagger) = \hat{a}_\alpha\hat{a}_\beta^\dagger + \hat{a}_\beta^\dagger\hat{a}_\alpha = \delta_{\alpha\beta}. \tag{3.1.6}$$

The first two contractions are zero since the operators are of the same type, and the normal ordering does not affect the order of such operators. The third term is zero because the initial string is already normal-ordered. The only contraction that is different from zero is the last one. Here, the operators were not in normal order from the start. Upon ordering them, a $-$ sign arises from the corresponding fermionic permutation and the final result is a Kronecker delta of the two states.

Clearly, the vacuum expectation values of the first three contractions is zero. For the fourth one, we find:

$$\langle 0|\overset{\frown}{\hat{a}_\alpha\hat{a}_\beta^\dagger}|0\rangle = \langle 0|\delta_{\alpha\beta}|0\rangle = \delta_{\alpha\beta}. \tag{3.1.7}$$

This expectation value is different from zero only when $\alpha = \beta$.

(b) By the Wick theorem, products of creation and destruction operators are formed of normal-ordered terms with no contractions; a pair of contractions; two pairs of contractions; and so on until all operators are contracted in pairs. The expectation value of any normal-ordered term is zero, because destruction operators are on the right and, upon acting on the vacuum, they yield 0. The only term that can be non-zero is the fully contracted one. However, if the number of operators is odd, is not possible to have all operators contracted in pairs. There will always be either a creation or a destruction operator left without a contraction. Upon evaluating the expectation value on the vacuum, the result is necessarily zero.

PROBLEM 3.2
Matrix elements using the Wick theorem: One-body operator

Consider the one-body operator

$$\hat{T} = \sum_{\alpha,\beta} (\alpha|\hat{t}|\beta)\hat{a}_\alpha^\dagger \hat{a}_\beta, \tag{3.2.1}$$

where the single-particle states α are the basis used to build the Fock space and the one-body operator in first quantization is

$$\hat{T} = \sum_{i=1}^{N} \hat{t}_i. \tag{3.2.2}$$

Consider the Fock states: $|c_1\rangle = \hat{a}_1^\dagger \hat{a}_2^\dagger |0\rangle$, $|c_2\rangle = \hat{a}_1^\dagger \hat{a}_4^\dagger |0\rangle$ and $|c_3\rangle = \hat{a}_4^\dagger \hat{a}_5^\dagger |0\rangle$. Calculate the following expectation values:

(a) $\langle c_1 | \hat{T} | c_1 \rangle$,

(b) $\langle c_1 | \hat{T} | c_2 \rangle$,

(c) $\langle c_1 | \hat{T} | c_3 \rangle$.

(a) We start with the diagonal matrix element,

$$\langle c_1 | \hat{T} | c_1 \rangle = \langle 0| \hat{a}_2 \hat{a}_1 \sum_{\alpha,\beta} (\alpha | \hat{t} | \beta)\hat{a}_\alpha^\dagger \hat{a}_\beta \hat{a}_1^\dagger \hat{a}_2^\dagger |0\rangle$$

$$= \sum_{\alpha,\beta} (\alpha | \hat{t} | \beta) \langle 0|\hat{a}_2\hat{a}_1\hat{a}_\alpha^\dagger \hat{a}_\beta \hat{a}_1^\dagger \hat{a}_2^\dagger|0\rangle. \tag{3.2.3}$$

Now, we should recall the results of the previous problem,

$$\overbracket{\hat{a}_\alpha \hat{a}_\beta} = 0, \quad \overbracket{\hat{a}_\alpha^\dagger \hat{a}_\beta^\dagger} = 0, \quad \overbracket{\hat{a}_\beta^\dagger \hat{a}_\beta} = 0, \quad \overbracket{\hat{a}_\alpha \hat{a}_\beta^\dagger} = \delta_{\alpha\beta}. \tag{3.2.4}$$

The only non-zero result of the expectation value is the one obtained from contracting all possible operators. We only consider contractions between different types of operators that are not normal ordered, which reduces to two possible fully-contracted terms:

$$\langle c_1 | \hat{T} | c_1 \rangle = \sum_{\alpha\beta} (\alpha | \hat{t} | \beta) \langle 0 | \hat{a}_2 \hat{a}_1 \hat{a}_\alpha^\dagger \hat{a}_\beta \hat{a}_1^\dagger \hat{a}_2^\dagger | 0 \rangle$$

$$+ \sum_{\alpha\beta} (\alpha | \hat{t} | \beta) \langle 0 | \hat{a}_2 \hat{a}_1 \hat{a}_\alpha^\dagger \hat{a}_\beta \hat{a}_1^\dagger \hat{a}_2^\dagger | 0 \rangle$$

$$= \sum_{\alpha\beta} (\alpha | \hat{t} | \beta) \left[\delta_{\alpha 1} \delta_{\beta 1} + \delta_{\alpha 2} \delta_{\beta 2} \right]$$

$$= (1 | \hat{t} | 1) + (2 | \hat{t} | 2). \tag{3.2.5}$$

In other words, the expectation value is the sum of the single-particle expectation values over the two single-particle states, 1 and 2, included in the $|c_1\rangle$ state.

(b) We now turn our attention to the calculation of a matrix element, which differs from the expectation value in that the bra and ket states are different. In this case, we find:

$$\langle c_1 | \hat{T} | c_2 \rangle = \sum_{\alpha\beta} (\alpha | \hat{t} | \beta) \langle 0 | a_2 a_1 a_\alpha^\dagger a_\beta a_1^\dagger a_4^\dagger | 0 \rangle$$

$$= \sum_{\alpha\beta} (\alpha | \hat{t} | \beta) \langle 0 | \hat{a}_2 \hat{a}_1 \hat{a}_\alpha^\dagger \hat{a}_\beta a_1^\dagger \hat{a}_4^\dagger | 0 \rangle$$

$$+ \sum_{\alpha\beta} (\alpha | \hat{t} | \beta) \langle 0 | \hat{a}_2 \hat{a}_1 \hat{a}_\alpha^\dagger \hat{a}_\beta \hat{a}_1^\dagger a_4^\dagger | 0 \rangle$$

$$= \sum_{\alpha\beta} (\alpha | \hat{t} | \beta) \delta_{2\alpha} \delta_{\beta 4}$$

$$= (2 | \hat{t} | 4). \tag{3.2.6}$$

In the end, the many-body matrix element is equal to one single-particle matrix elements between the two different states, 2 and 4. Clearly, this matrix element is non-zero because there is a contraction between the common state 1 included in both $|c_1\rangle$ and $|c_2\rangle$.

(c) We are finally asked to compute the matrix element $\langle c_1 | \hat{T} | c_3 \rangle$. In this case, $|c_1\rangle$ and $|c_3\rangle$ do not share any single-particle indices. Because the two states differ in all single-particle states and the operator is of one-body nature, there is no way to connect these two states with one another. As a consequence, the expectation value is null,

$$\langle c_1 | \hat{T} | c_3 \rangle = 0. \tag{3.2.7}$$

PROBLEM 3.3
Matrix elements using the Wick theorem: Two-body operator

Consider the two-body operator

$$\hat{V} = \frac{1}{2} \sum_{\alpha\beta\gamma\delta} (\alpha\beta \mid \hat{v} \mid \gamma\delta) \hat{a}_\alpha^\dagger \hat{a}_\beta^\dagger \hat{a}_\delta \hat{a}_\gamma, \tag{3.3.1}$$

where

$$(\alpha\beta \mid \hat{v} \mid \gamma\delta) = \int d\vec{r} \int d\vec{r}' \, \varphi_\alpha^*(\vec{r}) \varphi_\beta^*(\vec{r}') v(\mid \vec{r} - \vec{r}' \mid) \varphi_\gamma(\vec{r}) \varphi_\delta(\vec{r}') \tag{3.3.2}$$

are the two-body matrix elements expressed in real space, and the single-particle states α are the basis used to build the Fock space.

Consider the Fock states:

$$|A\rangle = \hat{a}_3^\dagger \hat{a}_2^\dagger \hat{a}_1^\dagger |0\rangle, \quad |B\rangle = \hat{a}_4^\dagger \hat{a}_2^\dagger \hat{a}_1^\dagger |0\rangle, \quad |C\rangle = \hat{a}_5^\dagger \hat{a}_4^\dagger \hat{a}_1^\dagger |0\rangle \text{ and } |D\rangle = \hat{a}_6^\dagger \hat{a}_5^\dagger \hat{a}_4^\dagger |0\rangle.$$

Calculate the following expectation values:

(a) $\langle A \mid \hat{V} \mid A \rangle$,

(b) $\langle A \mid \hat{V} \mid B \rangle$,

(c) $\langle A \mid \hat{V} \mid C \rangle$,

(d) $\langle A \mid \hat{V} \mid D \rangle$.

(a) We start with the diagonal matrix element,

$$\langle A \mid \hat{V} \mid A \rangle = \langle 0 | \hat{a}_3 \hat{a}_2 \hat{a}_1 \frac{1}{2} \sum_{\alpha\beta\gamma\delta} (\alpha\beta \mid \hat{v} \mid \gamma\delta) \hat{a}_\alpha^\dagger \hat{a}_\beta^\dagger \hat{a}_\delta \hat{a}_\gamma \hat{a}_3^\dagger \hat{a}_2^\dagger \hat{a}_1^\dagger |0\rangle$$

$$= \frac{1}{2} \sum_{\alpha\beta\gamma\delta} (\alpha\beta \mid \hat{v} \mid \gamma\delta) \langle 0 | \hat{a}_3 \hat{a}_2 \hat{a}_1 \hat{a}_\alpha^\dagger \hat{a}_\beta^\dagger \hat{a}_\delta \hat{a}_\gamma \hat{a}_3^\dagger \hat{a}_2^\dagger \hat{a}_1^\dagger |0\rangle \tag{3.3.3}$$

We employ once again the results of Problem 3.1,

$$\overline{\hat{a}_\alpha \hat{a}_\beta} = 0, \quad \overline{\hat{a}_\alpha^\dagger \hat{a}_\beta^\dagger} = 0, \quad \overline{\hat{a}_\beta^\dagger \hat{a}_\alpha} = 0, \quad \overline{\hat{a}_\alpha \hat{a}_\beta^\dagger} = \delta_{\alpha\beta}. \tag{3.3.4}$$

We also need to take into account the phases originated by the parity of the permutations required to put the contracted operators together. With this, we find 12 contractions that are not zero:

$$\langle A | \hat{V} | A \rangle = \frac{1}{2} \sum_{\alpha\beta\gamma\delta} (\alpha\beta | \hat{v} | \gamma\delta)$$

$$\times \left[\langle 0|\hat{a}_1\hat{a}_2\hat{a}_3\hat{a}_\alpha^\dagger\hat{a}_\beta^\dagger\hat{a}_\delta\hat{a}_\gamma\hat{a}_3^\dagger\hat{a}_2^\dagger\hat{a}_1^\dagger|0\rangle + \langle 0|\hat{a}_1\hat{a}_2\hat{a}_3\hat{a}_\alpha^\dagger\hat{a}_\beta^\dagger\hat{a}_\delta\hat{a}_\gamma\hat{a}_3^\dagger\hat{a}_2^\dagger\hat{a}_1^\dagger|0\rangle \right.$$

$$+ \langle 0|\hat{a}_1\hat{a}_2\hat{a}_3\hat{a}_\alpha^\dagger\hat{a}_\beta^\dagger\hat{a}_\delta\hat{a}_\gamma\hat{a}_3^\dagger\hat{a}_2^\dagger\hat{a}_1^\dagger|0\rangle + \langle 0|\hat{a}_1\hat{a}_2\hat{a}_3\hat{a}_\alpha^\dagger\hat{a}_\beta^\dagger\hat{a}_\delta\hat{a}_\gamma\hat{a}_3^\dagger\hat{a}_2^\dagger\hat{a}_1^\dagger|0\rangle$$

$$+ \langle 0|\hat{a}_1\hat{a}_2\hat{a}_3\hat{a}_\alpha^\dagger\hat{a}_\beta^\dagger\hat{a}_\delta\hat{a}_\gamma\hat{a}_3^\dagger\hat{a}_2^\dagger\hat{a}_1^\dagger|0\rangle + \langle 0|\hat{a}_1\hat{a}_2\hat{a}_3\hat{a}_\alpha^\dagger\hat{a}_\beta^\dagger\hat{a}_\delta\hat{a}_\gamma\hat{a}_3^\dagger\hat{a}_2^\dagger\hat{a}_1^\dagger|0\rangle$$

$$+ \langle 0|\hat{a}_1\hat{a}_2\hat{a}_3\hat{a}_\alpha^\dagger\hat{a}_\beta^\dagger\hat{a}_\delta\hat{a}_\gamma\hat{a}_3^\dagger\hat{a}_2^\dagger\hat{a}_1^\dagger|0\rangle + \langle 0|\hat{a}_1\hat{a}_2\hat{a}_3\hat{a}_\alpha^\dagger\hat{a}_\beta^\dagger\hat{a}_\delta\hat{a}_\gamma\hat{a}_3^\dagger\hat{a}_2^\dagger\hat{a}_1^\dagger|0\rangle$$

$$+ \langle 0|\hat{a}_1\hat{a}_2\hat{a}_3\hat{a}_\alpha^\dagger\hat{a}_\beta^\dagger\hat{a}_\delta\hat{a}_\gamma\hat{a}_3^\dagger\hat{a}_2^\dagger\hat{a}_1^\dagger|0\rangle + \langle 0|\hat{a}_1\hat{a}_2\hat{a}_3\hat{a}_\alpha^\dagger\hat{a}_\beta^\dagger\hat{a}_\delta\hat{a}_\gamma\hat{a}_3^\dagger\hat{a}_2^\dagger\hat{a}_1^\dagger|0\rangle$$

$$\left. + \langle 0|\hat{a}_1\hat{a}_2\hat{a}_3\hat{a}_\alpha^\dagger\hat{a}_\beta^\dagger\hat{a}_\delta\hat{a}_\gamma\hat{a}_3^\dagger\hat{a}_2^\dagger\hat{a}_1^\dagger|0\rangle + \langle 0|\hat{a}_1\hat{a}_2\hat{a}_3\hat{a}_\alpha^\dagger\hat{a}_\beta^\dagger\hat{a}_\delta\hat{a}_\gamma\hat{a}_3^\dagger\hat{a}_2^\dagger\hat{a}_1^\dagger|0\rangle \right].$$

$$(3.3.5)$$

The result of the contractions can be expressed in terms of Kronecker deltas,

$$\langle A | \hat{V} | A \rangle = \frac{1}{2} \sum_{\alpha\beta\gamma\delta} (\alpha\beta | \hat{v} | \gamma\delta)$$

$$\times \left[-\delta_{\alpha2}\delta_{\beta3}\delta_{\gamma3}\delta_{\delta2} + \delta_{\alpha2}\delta_{\beta3}\delta_{\gamma2}\delta_{\delta3} \right.$$

$$+ \delta_{\alpha3}\delta_{\beta2}\delta_{\gamma3}\delta_{\delta2} - \delta_{\alpha3}\delta_{\beta2}\delta_{\gamma2}\delta_{\delta3}$$

$$- \delta_{\alpha1}\delta_{\beta3}\delta_{\gamma3}\delta_{\delta1} + \delta_{\alpha1}\delta_{\beta3}\delta_{\gamma1}\delta_{\delta3}$$

$$+ \delta_{\alpha3}\delta_{\beta1}\delta_{\gamma3}\delta_{\delta1} - \delta_{\alpha3}\delta_{\beta1}\delta_{\gamma1}\delta_{\delta3}$$

$$- \delta_{\alpha1}\delta_{\beta2}\delta_{\gamma2}\delta_{\delta1} + \delta_{\alpha1}\delta_{\beta2}\delta_{\gamma1}\delta_{\delta2}$$

$$\left. + \delta_{\alpha2}\delta_{\beta1}\delta_{\gamma2}\delta_{\delta1} - \delta_{\alpha2}\delta_{\beta1}\delta_{\gamma1}\delta_{\delta2} \right].$$

$$(3.3.6)$$

Collapsing the sums over single-particle indices with the help of the Kronecker functions, and using the fact that $\hat{V}$ is a central potential such that $(\alpha\beta | \hat{v} | \gamma\delta) =$

$(\beta\alpha \mid \hat{v} \mid \delta\gamma)$, we get:

$$\langle A|\hat{V}|A\rangle = \frac{1}{2}[-(23 \mid \hat{v} \mid 32) + (23 \mid \hat{v} \mid 23)$$
$$+ (32 \mid \hat{v} \mid 32) - (32 \mid \hat{v} \mid 23)$$
$$- (13 \mid \hat{v} \mid 31) + (13 \mid \hat{v} \mid 13)$$
$$+ (31 \mid \hat{v} \mid 31) - (31 \mid \hat{v} \mid 13)$$
$$- (12 \mid \hat{v} \mid 21) + (12 \mid \hat{v} \mid 12)$$
$$+ (21 \mid \hat{v} \mid 21) - (21 \mid \hat{v} \mid 12)]$$
$$= (23 \mid \hat{v} \mid 23 - 32) + (13 \mid \hat{v} \mid 13 - 31) + (12 \mid \hat{v} \mid 12 - 21). \quad (3.3.7)$$

Each one of the terms on the r.h.s. has both a so-called direct (or Hartree) matrix element, $(\alpha\beta \mid \hat{v} \mid \alpha\beta)$, and an exchange (or Fock) term, $(\alpha\beta \mid \hat{v} \mid \beta\alpha)$, where the bra and ket states are exchanged. The result is the sum over the 3 possible combinations of the 3 available states, 12, 13 and 23.

(b) Let us now calculate the matrix element $\langle A|\hat{V}|B\rangle$. In analogy with the previous case, we can write the matrix element in terms of fully contracted terms,

$$\langle A|\hat{V}|B\rangle = \frac{1}{2} \sum_{\alpha\beta\gamma\delta} (\alpha\beta \mid \hat{v} \mid \gamma\delta)$$

$$\times \left[\langle 0|\hat{a}_1\hat{a}_2\hat{a}_3\hat{a}_\alpha^\dagger\hat{a}_\beta^\dagger\hat{a}_\delta\hat{a}_\gamma\hat{a}_4^\dagger\hat{a}_2^\dagger\hat{a}_1^\dagger|0\rangle + \langle 0|\hat{a}_1\hat{a}_2\hat{a}_3\hat{a}_\alpha^\dagger\hat{a}_\beta^\dagger\hat{a}_\delta\hat{a}_\gamma\hat{a}_4^\dagger\hat{a}_2^\dagger\hat{a}_1^\dagger|0\rangle \right.$$

$$+ \langle 0|\hat{a}_1\hat{a}_2\hat{a}_3\hat{a}_\alpha^\dagger\hat{a}_\beta^\dagger\hat{a}_\delta\hat{a}_\gamma\hat{a}_4^\dagger\hat{a}_2^\dagger\hat{a}_1^\dagger|0\rangle + \langle 0|\hat{a}_1\hat{a}_2\hat{a}_3\hat{a}_\alpha^\dagger\hat{a}_\beta^\dagger\hat{a}_\delta\hat{a}_\gamma\hat{a}_4^\dagger\hat{a}_2^\dagger\hat{a}_1^\dagger|0\rangle$$

$$+ \langle 0|\hat{a}_1\hat{a}_2\hat{a}_3\hat{a}_\alpha^\dagger\hat{a}_\beta^\dagger\hat{a}_\delta\hat{a}_\gamma\hat{a}_4^\dagger\hat{a}_2^\dagger\hat{a}_1^\dagger|0\rangle + \langle 0|\hat{a}_1\hat{a}_2\hat{a}_3\hat{a}_\alpha^\dagger\hat{a}_\beta^\dagger\hat{a}_\delta\hat{a}_\gamma\hat{a}_4^\dagger\hat{a}_2^\dagger\hat{a}_1^\dagger|0\rangle$$

$$\left. + \langle 0|\hat{a}_1\hat{a}_2\hat{a}_3\hat{a}_\alpha^\dagger\hat{a}_\beta^\dagger\hat{a}_\delta\hat{a}_\gamma\hat{a}_4^\dagger\hat{a}_2^\dagger\hat{a}_1^\dagger|0\rangle + \langle 0|\hat{a}_1\hat{a}_2\hat{a}_3\hat{a}_\alpha^\dagger\hat{a}_\beta^\dagger\hat{a}_\delta\hat{a}_\gamma\hat{a}_4^\dagger\hat{a}_2^\dagger\hat{a}_1^\dagger|0\rangle \right].$$

$$(3.3.8)$$

Notice that now we have 8 terms instead of the 12 terms appearing in $\langle A|\hat{V}|A\rangle$, because state $|B\rangle$ has only 2 matching particles with state $|A\rangle$. Then, taking into account the value of the contractions and the corresponding phases to put together the operators to be contracted, we obtain

$$\langle A | \hat{V} | B \rangle = \frac{1}{2} \sum_{\alpha\beta\gamma\delta} (\alpha\beta | \hat{v} | \gamma\delta)$$

$$\times \left[\delta_{\alpha2}\delta_{\beta3}\delta_{\gamma2}\delta_{\delta4} - \delta_{\alpha2}\delta_{\beta3}\delta_{\gamma4}\delta_{\delta2} \right.$$
$$- \delta_{\alpha3}\delta_{\beta2}\delta_{\gamma2}\delta_{\delta4} + \delta_{\alpha3}\delta_{\beta2}\delta_{\gamma4}\delta_{\delta2}$$
$$+ \delta_{\alpha1}\delta_{\beta3}\delta_{\gamma1}\delta_{\delta4} - \delta_{\alpha1}\delta_{\beta3}\delta_{\gamma4}\delta_{\delta1}$$
$$\left. - \delta_{\alpha3}\delta_{\beta1}\delta_{\gamma1}\delta_{\delta4} + \delta_{\alpha3}\delta_{\beta1}\delta_{\gamma4}\delta_{\delta1} \right]$$

$$= \frac{1}{2} \left[(23 | \hat{v} | 24) - (23 | \hat{v} | 42) + (32 | \hat{v} | 42) - (32 | \hat{v} | 24) \right.$$
$$\left. + (13 | \hat{v} | 14) - (31 | \hat{v} | 41) + (31 | \hat{v} | 41) - (31 | \hat{v} | 14) \right]$$

$$= \frac{1}{2} \left[(23 | \hat{v} | 24 - 42) + (32 | \hat{v} | 42 - 24) \right.$$
$$\left. + (13 | \hat{v} | 14 - 41) + (31 | \hat{v} | 41 - 14) \right]$$
$$= (23 | \hat{v} | 24 - 42) + (13 | \hat{v} | 14 - 41). \tag{3.3.9}$$

(c) The matrix element $\langle A | \hat{V} | C \rangle$ is somewhat simpler, because there is only one common state (state 1), which reduces the number of fully contracted terms to 4:

$$\langle A | \hat{V} | C \rangle = \frac{1}{2} \sum_{\alpha\beta\gamma\delta} (\alpha\beta | \hat{v} | \gamma\delta)$$

$$\times \left[\langle 0 | \hat{a}_1 \hat{a}_2 \hat{a}_3 \hat{a}_\alpha^\dagger \hat{a}_\beta^\dagger \hat{a}_\delta \hat{a}_\gamma \hat{a}_5^\dagger \hat{a}_4^\dagger \hat{a}_1^\dagger | 0 \rangle + \langle 0 | \hat{a}_1 \hat{a}_2 \hat{a}_3 \hat{a}_\alpha^\dagger \hat{a}_\beta^\dagger \hat{a}_\delta \hat{a}_\gamma \hat{a}_5^\dagger \hat{a}_4^\dagger \hat{a}_1^\dagger | 0 \rangle \right.$$

$$\left. + \langle 0 | \hat{a}_1 \hat{a}_2 \hat{a}_3 \hat{a}_\alpha^\dagger \hat{a}_\beta^\dagger \hat{a}_\delta \hat{a}_\gamma \hat{a}_5^\dagger \hat{a}_4^\dagger \hat{a}_1^\dagger | 0 \rangle + \langle 0 | \hat{a}_1 \hat{a}_2 \hat{a}_3 \hat{a}_\alpha^\dagger \hat{a}_\beta^\dagger \hat{a}_\delta \hat{a}_\gamma \hat{a}_5^\dagger \hat{a}_4^\dagger \hat{a}_1^\dagger | 0 \rangle \right].$$

$$\tag{3.3.10}$$

With the results of the contractions and the corresponding phases, the sums reduce to a single term,

$$\langle A | \hat{V} | C \rangle = \frac{1}{2} \sum_{\alpha\beta\gamma\delta} (\alpha\beta | \hat{v} | \gamma\delta) \left[\delta_{\alpha2}\delta_{\beta3}\delta_{\gamma4}\delta_{\delta5} - \delta_{\alpha2}\delta_{\beta3}\delta_{\gamma5}\delta_{\delta4} \right.$$
$$\left. - \delta_{\alpha3}\delta_{\beta2}\delta_{\gamma4}\delta_{\delta5} + \delta_{\alpha3}\delta_{\beta2}\delta_{\gamma5}\delta_{\delta4} \right]$$
$$= \frac{1}{2} \left[(23 | \hat{v} | 45) - (23 | \hat{v} | 54) + (32 | \hat{v} | 54) - (32 | \hat{v} | 45) \right]$$
$$= (23 | \hat{v} | 45 - 54). \tag{3.3.11}$$

(d) Finally, the last matrix element is

$$\langle A | \hat{V} | D \rangle = 0, \tag{3.3.12}$$

because states $|A\rangle$ and $|D\rangle$ differ in two or more single-particle states. They cannot be connected by a two-body operator, and hence the matrix element is necessarily zero.

4 Green's functions

Green's functions are a cornerstone of quantum mechanics and many-body physics, providing a systematic framework to describe the propagation of particles and excitations in a system. They encode essential information about the evolution of a quantum particle in space and time, serving as the foundation for understanding properties such as energy levels, spectral functions and system responses to external perturbations. Their broad applicability makes them invaluable for studying systems ranging from simple, non-interacting particles, to highly correlated many-body systems.

This chapter introduces Green's functions through a progression of problems aimed at building both conceptual and practical skills. Starting with the Green function of a free particle in one dimension, the readers will explore how these functions capture the dynamics of non-interacting systems. Problems involving Fourier transforms of propagators illustrate how Green's functions connect the time and frequency domains, highlighting their importance in quantum dynamics and spectral analysis. This chapter also covers advanced topics, such as the definition of Green's functions in terms of field operators, and demonstrates their use in deriving key results like the kinetic energy and the Koltun sum rule for the free Fermi sea. By working through these problems, the readers will gain an understanding of the insights Green's functions provide into quantum mechanics and their pivotal role in modern theoretical physics.

PROBLEM 4.1
The Green's function of a free particle in 1D

Calculate the propagator of a free particle,

$$G(\alpha,\beta;t) = \frac{-i}{\hbar}\left\langle \alpha \left| e^{-i\frac{\hat{H}t}{\hbar}} \right| \beta \right\rangle, \tag{4.1.1}$$

in one dimension in both the coordinate and the momentum representations.

The Hamiltonian and the Schrödinger equation of a free particle read

$$\hat{H} = \frac{\hat{p}^2}{2m}, \quad i\hbar\frac{\partial |\Psi(t)\rangle}{\partial t} = \hat{H}|\Psi(t)\rangle, \tag{4.1.2}$$

while the time evolution operator is given by

$$\hat{U}(t) = e^{-i\hat{H}t/\hbar}, \tag{4.1.3}$$

such that

$$|\Psi(t)\rangle = \hat{U}(t)|\Psi(t=0)\rangle. \tag{4.1.4}$$

The eigenvectors of the Hamiltonian normalized to the continuum are given by

$$\langle x|p\rangle = \frac{1}{\sqrt{2\pi\hbar}}e^{ipx/\hbar}, \quad \hat{H}|p\rangle = \frac{p^2}{2m}|p\rangle. \tag{4.1.5}$$

In momentum space, the expression for the single-particle propagator reads

$$G(p,p';t) = \frac{-i}{\hbar}\left\langle p \left| e^{-i\frac{\hat{H}t}{\hbar}} \right| p' \right\rangle = -\frac{i}{\hbar}\delta(p-p')e^{-\frac{i}{\hbar}\frac{p^2}{2m}t}, \tag{4.1.6}$$

where we have taken into account that the momentum eigenstates, $|p\rangle$, are the eigenfunctions of $\hat{H}$,

$$\hat{H}|p\rangle = \frac{p^2}{2m}|p\rangle, \tag{4.1.7}$$

and they are orthonormal, $\langle p|p'\rangle = \delta(p-p')$. Note that Eq. (4.1.6) is indeed the propagator in momentum space, and it is thus the answer to the question for the momentum representation.

The definition of the propagator in the space representation is

$$G(x,x';t) = \frac{-i}{\hbar}\left\langle x \left| e^{-i\frac{\hat{H}t}{\hbar}} \right| x' \right\rangle. \tag{4.1.8}$$

Using the closure relation in momentum space $\int dp \, |p\rangle \langle p| = 1$ two times in this equation, we find

$$G(x,x';t) = -\frac{i}{\hbar} \int dp \int dp' \, \langle x|p'\rangle \left\langle p \left| e^{-i\frac{\hat{H}t}{\hbar}} \right| p' \right\rangle \langle p|x'\rangle . \tag{4.1.9}$$

We can now incorporate the expression of Eq. (4.1.6) in the integrals to find

$$G(x,x';t) = -\frac{i}{\hbar} \frac{1}{2\pi\hbar} \int dp \int dp' e^{i\frac{p'x}{\hbar}} \delta(p-p') e^{-\frac{i}{\hbar}\frac{p^2}{2m}t} e^{-i\frac{px'}{\hbar}}$$

$$= -\frac{i}{\hbar} \frac{1}{2\pi\hbar} \int dp \, e^{i\frac{p(x-x')}{\hbar}} e^{-\frac{i}{\hbar}\frac{p^2}{2m}t} .$$

With this, we can identify,

$$G(x,x';t) = -\frac{i}{\hbar} \frac{1}{2\pi\hbar} \int dp \, e^{-i\frac{p(x'-x)}{\hbar}} e^{-\frac{i}{\hbar}\frac{p^2}{2m}t} . \tag{4.1.10}$$

The next step is to perform this integral. The strategy is to complete the square in the exponent, and take advantage of the result $\int_{-\infty}^{\infty} d\phi \, e^{-\phi^2} = \sqrt{\pi}$. In fact, we can write

$$-i\frac{p^2 t}{2m\hbar} - i\frac{p(x'-x)}{\hbar} = -\left[\sqrt{\frac{it}{2m\hbar}} p + i\frac{\sqrt{m}}{\sqrt{2it\hbar}} (x'-x) \right]^2 - \frac{m}{2it\hbar} (x'-x)^2 . \tag{4.1.11}$$

Using this expression in the integral, one finds:

$$\frac{1}{2\pi\hbar} \int_{-\infty}^{\infty} dp \, e^{-\frac{i}{\hbar}\frac{p^2}{2m}t} e^{-i\frac{p(x'-x)}{\hbar}} = \frac{1}{2\pi\hbar} \int_{-\infty}^{\infty} dp \, e^{-\left[\sqrt{\frac{it}{2m\hbar}} p + i\frac{\sqrt{m}}{\sqrt{2it\hbar}} (x'-x) \right]^2} e^{-\frac{m}{2it\hbar} (x'-x)^2} . \tag{4.1.12}$$

Now, one can perform the change of variables,

$$\xi = \sqrt{\frac{it}{2m\hbar}} p + i\frac{\sqrt{m}}{\sqrt{2it\hbar}} (x'-x) , \quad dp = d\xi \sqrt{\frac{2m\hbar}{it}} , \tag{4.1.13}$$

so the integral becomes

$$\frac{1}{2\pi\hbar} \int_{-\infty}^{\infty} dp \, e^{-\frac{i}{\hbar}\frac{p^2}{2m}t} e^{-i\frac{p(x'-x)}{\hbar}} = \frac{1}{2\pi\hbar} \sqrt{\frac{2m\hbar}{it}} e^{-\frac{m}{2it\hbar} (x'-x)^2} \int_{-\infty}^{\infty} d\xi \, e^{-\xi^2} , \tag{4.1.14}$$

where we have pulled out the momentum-independent term in the exponent. Finally, one writes the propagator in real space in 1D:

$$G(x,x';t) = -\frac{i}{\hbar} \frac{1}{2\pi\hbar} \sqrt{\frac{2m\hbar}{it}} e^{-\frac{m}{2it\hbar} (x'-x)^2} \int_{-\infty}^{\infty} d\xi \, e^{\xi^2}$$

$$= -\frac{i}{\hbar} \sqrt{\frac{m}{2\pi i\hbar t}} e^{i\frac{m}{2\hbar t} (x'-x)^2} . \tag{4.1.15}$$

This is an oscillatory function in terms of the difference $(x-x')^2$.

In 3D, the Green's function can be determined following the same procedure for each dimension (x, y and z). Taking the product of the three terms, we find the expression of the 3D propagator:

$$G(\vec{r}, \vec{r}'; t) = -\frac{i}{\hbar} \left(\frac{m}{2\pi i \hbar t} \right)^{3/2} e^{i\frac{m}{2t\hbar}(\vec{r}' - \vec{r})^2}. \qquad (4.1.16)$$

PROBLEM 4.2
Single-particle spectral function for a non-interacting system

Calculate the one-body Green's function,

$$g^{(0)}(\alpha, \beta; t, t') \equiv \frac{-i}{\hbar} \langle \Psi_0^N | T \left[\hat{a}_{\alpha_H}(t) \hat{a}_{\beta_H}^\dagger(t') \right] | \Psi_0^N \rangle, \qquad (4.2.1)$$

for a non-interacting system described by the Hamiltonian

$$\hat{H}_0 = \sum_\alpha \varepsilon_\alpha \hat{a}_\alpha^\dagger \hat{a}_\alpha, \qquad (4.2.2)$$

with all single-particle states occupied up to the Fermi level α_F. The ground state $|\Psi_0^N\rangle = \prod_{i \leq F} \hat{a}_i^\dagger |0\rangle$ is an eigenstate of $\hat{H}_0$,

$$\hat{H}_0 |\Psi_0^N\rangle = E_0 |\Psi_0^N\rangle, \quad E_0 = \sum_{\alpha \leq F} \varepsilon_\alpha. \qquad (4.2.3)$$

The time ordered single-particle Green's function is defined by the expectation value in Eq. (4.2.1), where $|\Psi_0^N\rangle$ is the ground state and all quantities are defined in the Heisenberg representation. The time ordering operator is given by

$$T \left[\hat{a}_{\alpha_H}(t) \hat{a}_{\beta_H}^\dagger(t') \right] = \Theta(t - t') \hat{a}_{\alpha_H}(t) \hat{a}_{\beta_H}^\dagger(t') - \Theta(t' - t) \hat{a}_{\beta_H}^\dagger(t') \hat{a}_{\alpha_H}(t), \quad (4.2.4)$$

where $\Theta(t)$ is the step function, which is 0 for $t < 0$ and 1 for $t > 0$. Notice that this definition includes a sign change when the two fermion operators are interchanged. The idea is that the time-ordering operator puts operators with earlier time argument to the right, so they act first on the corresponding kets.

In the definition, $|\Psi_0^N\rangle$ is the normalized Heisenberg-representation ground state. Finding this state is usually a problem of its own. However, full knowledge of the state itself is not necessary to evaluate the Green's function. The key point in the present case is that this ground state has a known energy, E_0, as per Eq. (4.2.3). In the Heisenberg picture, the single-particle operators $\hat{a}_{\alpha_H}(t)$ and $\hat{a}_{\alpha_H}^\dagger$ are given by

$$\hat{a}_{\alpha_H}(t) = e^{i\frac{\hat{H}t}{\hbar}} \hat{a}_\alpha e^{-i\frac{\hat{H}t}{\hbar}}, \qquad (4.2.5)$$

and

$$\hat{a}_{\alpha_H}^\dagger(t) = e^{i\frac{\hat{H}t}{\hbar}} \, \hat{a}_\alpha^\dagger \, e^{-i\frac{\hat{H}t}{\hbar}} . \tag{4.2.6}$$

Then, taking into account all these definitions, for any $\hat{H}$ we can write

$$g(\alpha,\beta;t-t') = \frac{-i}{\hbar} \left\{ \Theta(t-t') \, \langle \Psi_0^N | e^{i\frac{\hat{H}t}{\hbar}} \, \hat{a}_\alpha \, e^{-i\frac{\hat{H}t}{\hbar}} \, e^{i\frac{\hat{H}t'}{\hbar}} \, \hat{a}_\beta^\dagger \, e^{-i\frac{\hat{H}t'}{\hbar}} \, | \Psi_0^N \rangle \right.$$
$$\left. - \Theta(t'-t) \, \langle \Psi_0^N | \, e^{i\frac{\hat{H}t'}{\hbar}} \, \hat{a}_\beta^\dagger \, e^{-i\frac{\hat{H}t'}{\hbar}} \, e^{i\frac{\hat{H}t}{\hbar}} \, \hat{a}_\alpha \, e^{-i\frac{\hat{H}t}{\hbar}} \, | \Psi_0^N \rangle \right\} , \tag{4.2.7}$$

which can be simplified as

$$g(\alpha,\beta;t-t') = \frac{-i}{\hbar} \left\{ \Theta(t-t') e^{iE_0^N(t-t')/\hbar} \, \langle \Psi_0^N | \hat{a}_\alpha e^{-i\hat{H}(t-t')/\hbar} \hat{a}_\beta^\dagger | \Psi_0^N \rangle \right.$$
$$\left. - \Theta(t'-t) e^{iE_0^N(t'-t)\hbar} \, \langle \Psi_0^N | \hat{a}_\beta^\dagger e^{-i\hat{H}(t'-t)/\hbar} \hat{a}_\alpha | \Psi_0^n \rangle \right\} . \tag{4.2.8}$$

Using a closure relation for the eigenstates of the $N+1$ and the $N-1$ particle systems, we can write

$$g(\alpha,\beta;t-t') =$$
$$\frac{-i}{\hbar} \left\{ \Theta(t-t') \sum_m e^{i(E_0^N - E_m^{N+1})(t-t')/\hbar} \, \langle \Psi_0^N | \hat{a}_\alpha | \Psi_m^{N+1} \rangle \langle \Psi_m^{N+1} | \hat{a}_\beta^\dagger | \Psi_0^N \rangle \right.$$
$$\left. - \Theta(t'-t) \sum_{m'} e^{i(E_0^N - E_{m'}^{N-1})(t'-t)/\hbar} \, \langle \Psi_0^N | \hat{a}_\beta^\dagger | \Psi_{m'}^{N-1} \rangle \langle \Psi_{m'}^{N-1} | \hat{a}_\alpha | \Psi_0^N \rangle \right\} . \tag{4.2.9}$$

Now, we should take into account that for the overlap to be nonzero, the state $| \Psi_m^{N+1} \rangle$ must be constructed by just adding a particle to the state $| \Psi_0^N \rangle$. To add a particle, this must lie on a single-particle energy level above the Fermi level α_F, which means that

$$\langle \Psi_0^N | \hat{a}_\alpha | \Psi_m^{N+1} \rangle \, \langle \Psi_m^{N+1} | \hat{a}_\beta^\dagger \, | \Psi_0^N \rangle = \delta_{\alpha\beta} \, \Theta(\alpha - \alpha_F) \langle \Psi_m^{N+1} | \hat{a}_\alpha^\dagger | \Psi_0^N \rangle , \tag{4.2.10}$$

Similarly, for the state $| \Psi_m^{N-1} \rangle$ a particle is subtracted to the state $| \Psi_0^N \rangle$ from a single-particle energy level below the Fermi level

$$\langle \Psi_0^N | \hat{a}_\beta^\dagger | \Psi_m^{N-1} \rangle \, \langle \Psi_m^{N-1} | \hat{a}_\alpha | \Psi_0^N \rangle = \delta_{\alpha\beta} \, \Theta(\alpha_F - \alpha) \langle \Psi_m^{N-1} | \hat{a}_\alpha | \Psi_0^N \rangle . \tag{4.2.11}$$

Notice that in our case, i.e. the free Hamiltonian, we have $E_0 - E_m^{N+1} = -\varepsilon_\alpha$. The excitation energy of the $N+1$ particle state measured with respect to the ground state of N particles is just the single-particle energy of the state added to the system: ε_α. Similarly, for the non-interacting case, we know that $E_0^N - E_m^{N-1} = \varepsilon_\alpha$.

The $\langle \Psi_m^{N-1} | \hat{a}_\alpha | \Psi_0^N \rangle$, $\langle \Psi_m^{N+1} | \hat{a}_\alpha^\dagger | \Psi_0^N \rangle$ conditions permit to perform the summations and we get :

$$g(\alpha, \beta; t - t') = \frac{-i\delta_{\alpha\beta}}{\hbar} \left\{ \Theta(t - t')\, e^{-i\varepsilon_\alpha(t-t')/\hbar}\, \Theta(\alpha - \alpha_F) \right.$$
$$\left. - \Theta(t' - t)\, e^{i\varepsilon_\alpha(t'-t)/\hbar}\, \Theta(\alpha_F - \alpha) \right\}. \quad (4.2.12)$$

PROBLEM 4.3
Fourier transform of the single-particle propagator

Calculate the Fourier transform of the single-particle propagator

$$g^{(0)}(\alpha, \beta; E) = \int_{-\infty}^{\infty} d(t - t') e^{iE(t-t')/\hbar} g^{(0)}(\alpha, \beta; t - t'), \quad (4.3.1)$$

where

$$g^{(0)}(\alpha, \beta; t - t') = -\frac{i\delta_{\alpha\beta}}{\hbar} \left\{ \Theta(t - t') e^{-i\varepsilon_\alpha(t-t')/\hbar} \Theta(\alpha - \alpha_F) \right.$$
$$\left. - \Theta(t' - t) e^{i\varepsilon_\alpha(t'-t)/\hbar} \Theta(\alpha_F - \alpha) \right\} \quad (4.3.2)$$

is the one-body Green's function corresponding to the Hamiltonian

$$\hat{H}_o = \sum_\alpha \varepsilon_\alpha \hat{a}_\alpha^\dagger \hat{a}_\alpha \quad (4.3.3)$$

with all single-particle states occupied up to the Fermi level α_F, and the ground state $|\phi_0^N\rangle = \prod_{i < F} \hat{a}_i^\dagger |0\rangle$ is an eigenstate of $\hat{H}_0$

$$\hat{H}_0 |\phi_0^N\rangle = E_0 |\phi_0^n\rangle \quad , \quad E_0 = \sum_{\alpha \leq F} \varepsilon_\alpha \quad (4.3.4)$$

To calculate the Fourier transform, it is useful to consider the integral representation of the Θ-function:

$$\Theta(t - t') = -\frac{1}{2\pi i} \int_{-\infty}^{\infty} dE' \frac{e^{-iE'(t-t')/\hbar}}{E' + i\eta^+} \quad (4.3.5)$$

We start from the definition of the Fourier transform:

$$g^{(0)}(\alpha,\beta;E) = \int_{-\infty}^{\infty} d(\tau)e^{iE\tau/\hbar} \, g^{(0)}(\alpha,\beta;\tau)$$

$$= -i\int_{-\infty}^{\infty} d\left(\frac{\tau}{\hbar}\right) e^{iE\tau/\hbar} \left\{ -\frac{1}{2\pi i}\int_{-\infty}^{\infty} dE' \frac{e^{-iE'\tau/\hbar}}{E'+i\eta^+} \right\} \delta_{\alpha\beta}\Theta(\alpha-\alpha_F)e^{-i\varepsilon_\alpha\tau/\hbar}$$

$$+ i\int_{-\infty}^{\infty} d\left(\frac{\tau}{\hbar}\right) e^{iE\tau/\hbar} \left\{ -\frac{1}{2\pi i}\int_{-\infty}^{\infty} dE' \frac{e^{iE'\tau/\hbar}}{E'+i\eta^+} \right\} \delta_{\alpha\beta}\Theta(\alpha-\alpha_F)e^{-i\varepsilon_\alpha\tau/\hbar}.$$

$$(4.3.6)$$

where $\tau = t - t'$ has been used.

The next step, is to change the order of integration,

$$g^{(0)}(\alpha,\beta;E) = \int dE' \frac{1}{E'+i\eta^+}\delta_{\alpha\beta}\Theta(\alpha-\alpha_F)\frac{1}{2\pi}\int_{-\infty}^{\infty} d\left(\frac{\tau}{\hbar}\right) e^{i(E-E'-\varepsilon_\alpha)\tau/\hbar}$$

$$- \int dE' \frac{1}{E'+i\eta^+}\delta_{\alpha\beta}\Theta(\alpha_F-\alpha)\frac{1}{2\pi}\int_{-\infty}^{\infty} d\left(\frac{\tau}{\hbar}\right) e^{i(E+E'-\varepsilon_\alpha)\tau/\hbar}.$$

$$(4.3.7)$$

The integration over $d(\tau/\hbar)$ provides the δ-functions on the energy variables:

$$\frac{1}{2\pi}\int_{-\infty}^{\infty} d\left(\frac{\tau}{\hbar}\right) e^{i(E-E'-\varepsilon_\alpha)\tau/\hbar} = \delta(E-E'-\varepsilon_\alpha). \qquad (4.3.8)$$

$$\frac{1}{2\pi}\int_{-\infty}^{\infty} d\left(\frac{\tau}{\hbar}\right) e^{i(E+E'-\varepsilon_\alpha)\tau/\hbar} = \delta(E+E'-\varepsilon_\alpha). \qquad (4.3.9)$$

With the help of the δ-functions one can easily perform the integrations over the energy variable:

$$g^{(0)}(\alpha,\beta;E) = \int dE' \frac{1}{E'+i\eta^+}\delta_{\alpha\beta}\Theta(\alpha-\alpha_F)\delta(E-E'-\varepsilon_\alpha)$$

$$- \int dE' \frac{1}{E'+i\eta^+}\delta_{\alpha\beta}\Theta(\alpha_F-\alpha)\delta(E+E'-\varepsilon_\alpha)$$

$$= \delta_{\alpha\beta}\left[\Theta(\alpha-\alpha_F)\frac{1}{E-\varepsilon_\alpha+i\eta^+} - \Theta(\alpha_F-\alpha)\frac{1}{\varepsilon_\alpha-E+i\eta^+} \right]$$

$$= \delta_{\alpha\beta}\left[\frac{\Theta(\alpha-\alpha_F)}{E-\varepsilon_\alpha+i\eta^+} + \frac{\Theta(\alpha_F-\alpha)}{E-\varepsilon_\alpha-i\eta^+} \right]. \qquad (4.3.10)$$

Notice that one can also arrive at this expression by starting from the definition

$$g^{(0)}(\alpha,\beta;E) = \langle \Psi_0^N | a_\alpha \frac{1}{E-(\hat{H}_0-E_{\Psi_0^N})+i\eta^+} a_\beta^\dagger |\Psi_0^N\rangle$$

$$+ \langle \Psi_0^N | a_\beta^\dagger \frac{1}{E-(E_{\Psi_0^N}-\hat{H}_0)-i\eta^+} a_\alpha |\Psi_0^N\rangle. \qquad (4.3.11)$$

In our case,

$$\hat{H}_0 a_\beta^\dagger |\Psi_0^N\rangle = (E_{\Psi_0^N} + \varepsilon_\beta) a_\beta^\dagger |\Psi_0^N\rangle , \tag{4.3.12}$$

and

$$\hat{H}_0 a_\beta |\Psi_0^N\rangle = (E_{\Psi_0^N} - \varepsilon_\beta) a_\beta |\Psi_0^N\rangle . \tag{4.3.13}$$

Finally,

$$
g^{(0)}(\alpha,\beta;E) = \delta_{\alpha\beta} \left[\frac{\Theta(\alpha - \alpha_F)}{E - (E_{\Psi_0^N} + \varepsilon_\alpha - E_{\Psi_0^N}) + i\eta^+} \right.
$$
$$
\left. + \frac{\Theta(\alpha_F - \alpha)}{E - (E_{\Psi_0^N} - (E_{\Psi_0^N} - \varepsilon_\alpha)) - i\eta^+} \right]
$$
$$
= \delta_{\alpha\beta} \left[\frac{\Theta(\alpha - \alpha_F)}{E - \varepsilon_\alpha + i\eta^+} + \frac{\Theta(\alpha_F - \alpha)}{E - \varepsilon_\alpha - i\eta^+} \right]. \tag{4.3.14}
$$

PROBLEM 4.4
Green's function. Field operator definition.

Calculate the one-body Green's function for the free Fermi sea, using the definition in terms of field operators:

$$G_{\sigma\sigma'}(\vec{r}t,\vec{r}'t') = -\frac{i}{\hbar} \langle \phi_0 | T \left[\hat{\Psi}_{H,\sigma}(\vec{r},t) \hat{\Psi}_{H,\sigma'}^\dagger(\vec{r}'t') \right] | \phi_0 \rangle \tag{4.4.1}$$

where $|\phi_0\rangle = \prod_{k<k_F} \hat{a}_k^\dagger |0\rangle$ with $k \equiv \{\vec{k}, \sigma\}$, is the ground state of the free Fermi sea in the Fock space.

The field operators in the Heisenberg picture are given by

$$\hat{\Psi}_H(\vec{r},t) = \sum_m \varphi_m(\vec{r}) \, e^{-iE_m t/\hbar} \hat{a}_m , \tag{4.4.2}$$

where $\hat{a}_m$ is the annihilation operator associated to the single-particle state whose wavefunction is

$$\varphi(\vec{r}) = \frac{1}{\Omega^{1/2}} e^{i\vec{k}_m \vec{r}} , \tag{4.4.3}$$

with energy $E_m = \hbar^2 k_m^2/2m$. Introducing the definition of the time ordering operator we can write:

$$
\begin{aligned}
G(\vec{r}t;\vec{r}'t') = & -\frac{i}{\hbar}\Big(\langle\phi_0|\hat{\Psi}_H(\vec{r},t)\hat{\Psi}_H^\dagger(\vec{r}',t')|\phi_0\rangle\,\theta(t-t') \\
& -\langle\phi_0|\hat{\Psi}_H^\dagger(\vec{r}',t')\hat{\Psi}_H(\vec{r},t)|\phi_0\rangle\,\theta(t'-t)\Big) \\
= & -\frac{i}{\hbar}\sum_{lm}\varphi_l(\vec{r})\varphi_m^*(\vec{r}')\,e^{-i(E_l t - E_m t')/\hbar}\,\langle\phi_0|\hat{a}_l\hat{a}_m^\dagger|\phi_0\rangle\,\theta(t-t') \\
& -\frac{i}{\hbar}(-)\sum_{lm}\varphi_l(\vec{r})\varphi_m^*(\vec{r}')\,e^{-i(E_l t - E_m t')/\hbar}\,\langle\phi_0|\hat{a}_m^\dagger\hat{a}_l|\phi_0\rangle\,\theta(t'-t).
\end{aligned}
$$

$$(4.4.4)$$

For $t > t'$ we need to evaluate $\langle\phi_0|\hat{a}_l\hat{a}_m^\dagger|\phi_0\rangle = \theta(m-k_F)\,\delta_{lm}$ and for $t < t'$ we calculate $\langle\phi_0|\hat{a}_m^\dagger\hat{a}_l|\phi_0\rangle = \theta(k_F - l)\,\delta_{lm}$.

Then, the presence of the δ-functions allows us to perform part of the summations. Introducing also the step function in the summation indices we get

$$
\begin{aligned}
G(\vec{r}t;\vec{r}'t') = & \frac{-i}{\hbar}\sum_{m>k_F}\varphi_m(\vec{r})\varphi_m^*(\vec{r}')\,e^{-iE_m(t-t')/\hbar}\theta(t-t') \\
& +\frac{i}{\hbar}\sum_{m<k_F}\varphi_m(\vec{r})\varphi_m^*(\vec{r}')e^{-iE_m(t-t')/\hbar}\theta(t'-t).
\end{aligned}
$$

$$(4.4.5)$$

Now introducing again the spin-indexes, transforming the summation over momenta in an integral, and taking into account that the single-particle wavefunctions are plane waves, we arrive at

$$
\begin{aligned}
G_{\sigma\sigma'}(\vec{r}t;\vec{r}'t') = & \delta_{\sigma\sigma'}\frac{1}{(2\pi)^3}\frac{-i}{\hbar}\bigg(\int d\vec{k}\,e^{i\vec{k}(\vec{r}-\vec{r}')}\,e^{-iE_k(t-t')/\hbar}\theta(t-t')\,\theta(k-k_F) \\
& -\int d\vec{k}\,e^{i\vec{k}(\vec{r}-\vec{r}')}\,e^{-iE_k(t-t')/\hbar}\theta(t'-t)\,\theta(k_F-k)\bigg),
\end{aligned}
$$

$$(4.4.6)$$

with $E_k = \hbar^2 k^2/2m$.

PROBLEM 4.5
Single-particle spectral functions for a non-interacting system

Consider the single-particle Green's function of the free Fermi sea

$$g^{(0)}(k,E) = \frac{\Theta(k-k_F)}{E - \frac{\hbar^2 k^2}{2m} + i\eta} + \frac{\Theta(k_F-k)}{E - \frac{\hbar^2 k^2}{2m} - i\eta}.$$

$$(4.5.1)$$

(a) Calculate the single-particle spectral functions

(b) Calculate the occupation and the disoccupation of the single-particle states.

(a) To calculate the single-particle spectral function associated to $g^{(0)}(k,E)$ one should identify the imaginary part of $g^{(0)}(k,E)$. To this end, one should take into account that

$$\frac{1}{A \pm i\eta} = \frac{A \mp i\eta}{A^2 + \eta^2} = \frac{A}{A^2 + \eta^2} \mp i\frac{\eta}{A^2 + \eta^2} \tag{4.5.2}$$

and in the limit $\eta \to 0$

$$\lim_{\eta \to 0} \frac{1}{A \pm i\eta} = \frac{1}{A} \mp i\pi\delta(A), \tag{4.5.3}$$

where the delta function appears following its representation with functional series $\delta(x) = \lim_{\tau \to \infty} \frac{1}{\pi} \left(\frac{\tau}{\tau^2 x^2 + 1} \right) = \lim_{\eta \to 0} \frac{1}{\pi} \left(\frac{\eta}{x^2 + \eta^2} \right)$.

The hole part of the spectral function is defined as:

$$S_h(k,E) = \frac{1}{\pi} \text{Im}\, g^{(0)}(k,E), \quad E < \varepsilon_F. \tag{4.5.4}$$

Using the previous expression we have that

$$S_h(k,E) = \delta\left(E - \frac{\hbar^2 k^2}{2m} \right) \Theta(k_F - k), \quad E < \varepsilon_F \tag{4.5.5}$$

and for the particle part, we have

$$S_p(k,E) = -\frac{1}{\pi} \text{Im}\, g^{(0)}(k,E) = \delta\left(E - \frac{\hbar^2 k^2}{2m} \right) \Theta(k - k_F) \quad E > \varepsilon_F. \tag{4.5.6}$$

(b) In this case, the occupation coincides with the momentum distribution

$$n(k) = \int_{-\infty}^{\varepsilon_F} S_h(k,E)dE = \Theta(k_F - k) \int_{-\infty}^{\varepsilon_F} \delta\left(E - \frac{\hbar^2 k^2}{2m} \right) dE = \Theta(k_F - k) \tag{4.5.7}$$

The disoccupation will be given by :

$$d(k) = \int_{\varepsilon_F}^{\infty} S_p(k,E)dE = \Theta(k - k_F). \tag{4.5.8}$$

Notice that

$$n(k) + d(k) = 1 \tag{4.5.9}$$

i.e., the occupation plus the disoccupation of each single-particle state is 1.

> **PROBLEM 4.6**
> **Kinetic energy and Koltun sum-rule for the free Fermi sea.**
>
> Using the hole part of the single-particle spectral function of the free Fermi sea,
>
> $$S_h(k,E) = \delta \left(E - \frac{\hbar^2 k^2}{2m} \right) \Theta(k_F - k), \qquad (4.6.1)$$
>
> (a) Calculate the expectation value of the kinetic energy per particle in the free Fermi sea.
>
> (b) Using the Koltun sum-rule, calculate the total energy (kinetic energy) of the free Fermi sea.

(a) The expectation value of the kinetic energy per particle of a uniform system can be expressed as

$$\frac{1}{N} \langle \hat{T} \rangle = \frac{1}{N} \frac{\Omega}{(2\pi)^3} v \int d\vec{k} \, \langle \vec{k} | \hat{t} | \vec{k} \rangle \int_{-\infty}^{\varepsilon_F} S_h(k,E) \, dE, \qquad (4.6.2)$$

where Ω is the volume enclosing the system, N is the number of particles, v is the spin-isospin degeneracy and the one-body matrix element $\langle \vec{k} | \hat{t} | \vec{k} \rangle$ is given by

$$\langle \vec{k} | \hat{t} | \vec{k} \rangle = \frac{\hbar^2 k^2}{2m}, \qquad (4.6.3)$$

and the hole part of the spectral function is

$$S_h(k,E) = \delta \left(E - \frac{\hbar^2 k^2}{2m} \right) \Theta(k_F - k), \qquad (4.6.4)$$

Then,

$$\int_{-\infty}^{\varepsilon_F} S_h(k,E) \, dE = \Theta(k_F - k) \qquad (4.6.5)$$

and therefore

$$\begin{aligned}
\frac{1}{N} \langle \hat{T} \rangle &= \frac{1}{\rho} \frac{v}{(2\pi)^3} \int d\vec{k} \, \frac{\hbar^2 k^2}{2m} \Theta(k_F - k) \\
&= \frac{1}{\rho} \frac{v}{(2\pi)^3} \int_0^{k_F} dk \, \frac{\hbar^2 k^4}{2m} \int_0^{\pi} d\theta \sin(\theta) \int_0^{2\pi} d\varphi \qquad (4.6.6) \\
&= \frac{1}{\rho} \frac{v}{(2\pi)^3} \frac{1}{5} \frac{\hbar^2 k_F^5}{2m} 4\pi = \frac{3}{5} \frac{\hbar^2 k_F^2}{2m}.
\end{aligned}$$

where $\rho = N/\Omega$ is the density of the system and $k_F^3 = 6\pi^2 \rho / v$.

(b) The energy per particle in the ground state of a uniform system interacting through a two-body potential can be expressed in terms of the hole part of the spectral function by means of the Koltun sum-rule:

$$\frac{1}{N}\langle \hat{H}\rangle = \frac{1}{\rho}\frac{\nu}{(2\pi)^3}\frac{1}{2}\int d\vec{k}\int_{-\infty}^{\varepsilon_F}\left(\langle \vec{k}|\hat{t}|\vec{k}\rangle + E\right)S_h(k,E)dE. \tag{4.6.7}$$

Taking into account the previous calculations and the expression of $S_h(k,E)$ for the free Fermi sea, we have

$$\begin{aligned}\frac{1}{N}\langle \hat{H}\rangle_{FS} &= \frac{1}{\rho}\frac{\nu}{(2\pi)^3}\frac{1}{2}\int d\vec{k}\int_{-\infty}^{\varepsilon_F}\left(\frac{\hbar^2k^2}{2m}+E\right)S_h(k,E)dE\\ &= \frac{1}{\rho}\frac{\nu}{(2\pi)^3}\frac{1}{2}\int_{|\vec{k}|\leq k_F}d\vec{k}\,2\,\frac{\hbar^2k^2}{2m} = \frac{3}{5}\frac{\hbar^2k_F^2}{m} = 2\frac{1}{N}\langle \hat{T}\rangle\end{aligned} \tag{4.6.8}$$

Section II

Applications

5 The Fermi sea in one dimension

Most of the visible matter around us is made of fermions, e.g. neutrons, protons and electrons. Their properties are thus relevant to understand the behavior of matter in different situations and at very different scales. Fermions, in contrast to bosons, are described by antisymmetric wavefunctions, as explained in Chapter 1.3. This has an important consequence already for non-interacting fermionic systems. Namely, the way a set of non-interacting identical fermions builds its ground state is quite peculiar as compared to bosons and gives rise to the so-called Fermi sea.

In general, fermionic gases appear in nature in three-dimensional configurations, namely, building the electronic cloud of a nucleus, the electronic gas of a metal, or even the interior of a neutron star. Understanding the properties of fermionic gases provides a consistent explanation to phenomena in a wide range of scales appearing in many fields of physics. A prominent example is the fact that a free Fermi sea exerts a certain pressure. This stems directly from the Pauli exclusion principle, and far from being an academic boutade, it has direct consequences in seemingly uncorrelated physical systems, such as the stability of white dwarves in an astrophysical context or that of atomic nuclei in nuclear physics.

Up to very recently, we had at hand a variety of Fermi seas in nature that we could study, e.g. different nuclei, metals, neutron stars, etc. This situation has changed recently, with ultracold atomic gases experiments, where fermionic systems can be prepared in an extremely controlled way [13], allowing also to extract thermodynamic properties of custom-made fermionic gases [10].

Low dimensional Fermi gases are also of great interest today, again triggered by the fast advances in experimental methods to produce one- and two-dimensional degenerate quantum gases. The one dimensional case is particularly interesting, as the quantum effects are in general enhanced [8].

In this chapter we present the main tools to compute the properties of one-dimensional Fermi seas, exploring the equation of state, the pressure, etc. We also discuss the correlations appearing between the fermions in the sea. These correlations, which build from the Pauli principle, are fairly universal and thus important for a variety of systems. Finally, we consider the case of trapped fermions, which can in principle be produced in ultracold atomic gases.

PROBLEM 5.1
Thermodynamic properties of the free Fermi sea in 1D

Consider a uniform system of non-interacting spin-1/2 fermions in one dimension at zero temperature. You can consider the system as confined in a line of length L with periodic boundary conditions.

(a) Find the relation between the density, ρ, and the Fermi momentum, k_F.

(b) Calculate the total energy of the system as a function of ρ.

(c) Calculate the chemical potential, μ.

(d) Calculate the pressure exerted by these particles.

(e) Calculate the incompressibility, k_T^{-1}.

(f) Calculate the speed of sound.

(a) The volume occupied in configuration space for each quantum state is given by $2\pi/L$ where L is the length of the box on which we will impose periodic boundary conditions. Then,

$$N = \frac{L\,\nu}{2\pi} \int_{-k_F}^{k_F} dk = \frac{\nu\,L}{2\pi} 2k_F \,, \tag{5.1.1}$$

where ν is the spin degeneracy, $\nu = 2$ for spin 1/2. Then, we can write

$$\rho = \frac{N}{L} = \frac{\nu\,k_F}{\pi} \,. \tag{5.1.2}$$

In one dimension, the density is linear with the Fermi momentum.

(b) To calculate the total energy, we must sum up the kinetic energy of all occupied single-particle states:

$$E = \frac{L\,\nu}{2\pi} \int_{-k_F}^{k_F} dk \frac{\hbar^2 k^2}{2m} = \frac{L\,\nu}{2\pi} \frac{\hbar^2}{2m} 2 \frac{k_F^3}{3} \,. \tag{5.1.3}$$

Now, if we identify $N = L\,\nu\,k_F/\pi$, we can write

$$E = N \frac{1}{3} \frac{\hbar^2 k_F^2}{2m} \,, \tag{5.1.4}$$

and the energy per particle as

$$e = \frac{E}{N} = \frac{1}{3} \frac{\hbar^2 k_F^2}{2m} \,. \tag{5.1.5}$$

The total energy per particle is one-third of the Fermi energy. The total energy per particle can also be expressed in terms of the density as:

$$e(\rho) = \frac{1}{3}\frac{\hbar^2}{2m}\frac{\pi^2\rho^2}{v^2} . \tag{5.1.6}$$

The total energy per particle scales as the density squared.

(c) For the chemical potential we should calculate:

$$\mu = \left(\frac{\partial E}{\partial N}\right)_L = \left(\frac{\partial(e(\rho)N)}{\partial N}\right)_L = e(\rho) + \rho\frac{\partial e(\rho)}{\partial \rho} , \tag{5.1.7}$$

that in our case is

$$\mu(\rho) = \frac{1}{3}\frac{\hbar^2}{2m}\frac{\pi^2\rho^2}{v^2} + \frac{2}{3}\frac{\hbar^2}{2m}\frac{\pi^2\rho^2}{v^2} = \frac{\hbar^2}{2m}\frac{\pi^2\rho^2}{v^2} . \tag{5.1.8}$$

Now, if one recognizes $k_F^2 = \pi^2\rho^2/v^2$, then

$$\mu = \frac{\hbar^2 k_F^2}{2m} . \tag{5.1.9}$$

The chemical potential is the energy of the last occupied single-particle state.

(d) The pressure is given by

$$P = -\left(\frac{\partial E}{\partial L}\right)_N = -N\frac{\partial e(\rho)}{\partial \rho}\frac{\partial \rho}{\partial L} = \rho^2\frac{\partial e(\rho)}{\partial \rho} . \tag{5.1.10}$$

Then,

$$P = \rho^2\frac{\partial e(\rho)}{\partial \rho} = \frac{1}{3}\frac{\hbar^2}{2m}\frac{2\pi^2}{v^2}\rho^3 . \tag{5.1.11}$$

(e) To calculate the incompressibility,

$$k_T^{-1} = \rho\frac{\partial P(\rho)}{\partial \rho} = 2\rho^2\frac{\partial e(\rho)}{\partial \rho} + \rho^3\frac{\partial^2 e(\rho)}{\partial \rho^2} , \tag{5.1.12}$$

which in our case becomes,

$$k_T^{-1} = 2\rho^3\frac{\hbar^2}{2m}\frac{\pi^2}{v^2} . \tag{5.1.13}$$

(f) Finally, the speed of sound:

$$c_s^2 = \frac{k_T^{-1}}{\rho m} = 2\rho^2\frac{\hbar^2}{2m^2}\frac{\pi^2}{v^2} = 2\frac{v^2 k_F^2}{\pi^2}\frac{\hbar^2}{2m^2}\frac{\pi^2}{v^2} = \frac{\hbar^2 k_F^2}{m^2} = v_F^2 . \tag{5.1.14}$$

The speed of sound coincides with the speed associated to the Fermi momentum.

PROBLEM 5.2
Two-body distribution function of a free Fermi sea in 1D

Consider a uniform system of non-interacting spin-1/2 fermions in one dimension
at zero temperature. One can take a box of length L with periodic boundary conditions. The mean value per particle of a two-body operator $\hat{V} = \sum_{i<j} v(x_{ij})$ in a
uniform system is given by

$$\langle \phi_{\mathrm{gs}} | \hat{V} | \phi_{\mathrm{gs}} \rangle = \frac{1}{2} \rho \int dx_{12}\, v(x_{12})\, g(|x_{12}|), \tag{5.2.1}$$

where $g(|x_{12}|)$ is the two-body distribution function that depends only on the dis-
tance between two particles, and $x_{12} = x_2 - x_1$. The distribution function measures
the probability to find two particles at a given distance $|x_{12}|$, normalized such that
$\lim_{x_{12} \to \infty} g(|x_{12}|) \to 1$.

   Calculate the two-body distribution function of a uniform system of non-
interacting fermions of spin 1/2.

In the case of the free Fermi sea, the wavefunction of the ground state is a Slater
determinant built with all single-particle states filled up to the Fermi level. In this
case, the expectation value of $\hat{V}$ can be calculated as the sum of all two-body
matrix elements,

$$\langle \Phi_{\mathrm{FS}} | \hat{V} | \Phi_{\mathrm{FS}} \rangle = \frac{1}{2} \sum_{ij} \langle ij|\hat{v}|ij - ji \rangle, \tag{5.2.2}$$

where the single-particle states are plane waves,

$$\varphi_j(x) = \frac{1}{L^{1/2}} e^{ikx} s_j, \tag{5.2.3}$$

where $k_n = 2\pi n/L$, $x \in [-L/2, L/2]$, $n = 0, \pm 1, \pm 2, \ldots$ are the allowed momenta
when imposing periodic boundary conditions, and s_i represents the spin projection on the z-axis which can be up ($\uparrow$) or down ($\downarrow$). Now let's perform the summation of the two-body matrix elements:

$$\begin{aligned}
\frac{\langle \Phi_{\mathrm{FS}} | \hat{V} | \Phi_{\mathrm{FS}} \rangle}{N} &= \frac{1}{2} \frac{1}{N} \frac{L^2}{(2\pi)^2} \int_{-k_F}^{k_F} dk \int_{-k_F}^{k_F} dk' \\
&\times \sum_{s_1 s_2} \langle ks_1, k's_2 | \hat{v}(x_2 - x_1) | ks_1, k's_2 - k's_1, ks_1 \rangle.
\end{aligned} \tag{5.2.4}$$

We now introduce the single-particle wavefunctions in the two-body matrix elements and indicate the integration over the space coordinates:

$$
\frac{\langle \Phi_{FS} | \hat{V} | \Phi_{FS} \rangle}{N} = \frac{1}{2} \frac{1}{N} \frac{L^2}{(2\pi)^2} \int_{-k_F}^{k_F} dk \int_{-k_F}^{k_F} dk'
$$

$$
\times \sum_{s_1 s_2} \frac{1}{L^2} \int dx_1 \int dx_2 \, e^{-ikx_1} s_1(1) e^{-ik'x_2} s_2(2) \, v(x_2 - x_1)
$$

$$
\times \left[e^{ikx_1} s_1(1) e^{ik'x_2} s_2(2) - e^{ik'x_1} s_2(1) e^{ikx_2} s_1(2) \right]
$$

$$
= \frac{1}{2} \frac{1}{N} \frac{1}{(2\pi)^2} \int_{-k_F}^{k_F} dk \int_{-k_F}^{k_F} dk' \int dx_1 \int dx_2 \, v(x_2 - x_1)
$$

$$
\times \left[\sum_{s_1 s_2} 1 - \sum_{s_1 s_2} \delta_{s_1 s_2} e^{i(k-k')(x_2 - x_1)} \right]
$$

$$
= \frac{1}{2} \frac{1}{N} \frac{1}{(2\pi)^2} v^2 \int dx_1 \int dx_2 \, v(x_2 - x_1) \int_{-k_F}^{k_F} dk \int_{-k_F}^{k_F} dk'
$$

$$
\times \left(1 - \frac{1}{v} e^{ik(x_2 - x_1)} e^{-ik'(x_2 - x_1)} \right), \tag{5.2.5}
$$

where we have used that

$$
\sum_{s_1 s_2} 1 = v^2 \quad , \quad \sum_{s_1 s_2} \delta_{s_1 s_2} = v \tag{5.2.6}
$$

and we have exchanged the order of integration of coordinates and momenta. Now we evaluate the following integrals:

$$
\int_{-k_F}^{k_F} dk = 2k_F = (2\pi) \frac{\rho}{v}, \tag{5.2.7}
$$

$$
\int_{-k_F}^{k_F} dk \, e^{ikx} = \int_{-k_F}^{k_F} dk \, (\cos(kx) + i\sin(kx)) \tag{5.2.8}
$$

$$
= \int_{-k_F}^{k_F} dk \, \cos(kx) = \frac{\sin kx}{x} \Big|_{-k_F}^{k_F} = \frac{2\sin(k_F x)}{x},
$$

and introduce the definition of the Slater function in 1D:

$$
l_{1D}(k_F x) \equiv \frac{v}{(2\pi)\rho} \int_{-k_F}^{k_F} dk \, e^{ikx} = \frac{v}{(2\pi)\rho} 2 \frac{\sin(k_F x)}{x}. \tag{5.2.9}
$$

Then we can write:

$$
\int_{-k_F}^{k_F} dk \, e^{ikx} = \int_{-k_F}^{k_F} dk \, e^{-ikx} = (2\pi) \frac{\rho}{v} l_{1D}(k_F x). \tag{5.2.10}
$$

Therefore,

$$
\begin{aligned}
\frac{\langle \Phi_{FS} | \hat{V} | \Phi_{FS} \rangle}{N} &= \frac{1}{2} \frac{1}{N} \frac{1}{(2\pi)^2} v^2 \left[(2\pi)^2 \frac{\rho^2}{v^2} \iint dx_1 dx_2 v(x_{12}) \right. \\
&\left. \times \left(1 - \frac{l_{1D}^2(k_F x_{12})}{v} \right) \right] \\
&= \frac{1}{2} \frac{1}{N} \rho^2 \iint dx_1 dx_2 v(x_{12}) \left(1 - \frac{l_{1D}^2(k_F x_{12})}{v} \right). \qquad (5.2.11)
\end{aligned}
$$

Since the integrand depends solely on the relative coordinate x_{12}, it is convenient to perform the following coordinate transformation:

$$
x_{12} = x_2 - x_1, \quad x_{CM} = \frac{x_2 + x_1}{2},
$$

where x_{CM} is the position of the center of mass of the system. One can easily check that the Jacobian of this transformation is $+1$, and we thus have:

$$
\begin{aligned}
\frac{\langle \Phi_{FS} | \hat{V} | \Phi_{FS} \rangle}{N} &= \frac{1}{2} \frac{1}{N} \rho^2 \iint dx_{CM} dx_{12} v(x_{12}) \left(1 - \frac{l_{1D}^2(k_F x_{12})}{v} \right) \\
&= \frac{1}{2} \frac{1}{N} \rho^2 L \int dx_{12} v(x_{12}) \left(1 - \frac{l_{1D}^2(k_F x_{12})}{v} \right) \\
&= \frac{1}{2} \rho \int dx_{12} v(x_{12}) \left(1 - \frac{l_{1D}^2(k_F x_{12})}{v} \right),
\end{aligned}
$$

where we have used that $\int_{-L/2}^{L/2} dx_{CM} = L$. We can now identify the two-body distribution function as:

$$
g(x_{12}) = 1 - \frac{l_{1D}^2(k_F x_{12})}{v} \quad , \quad l_{1D}(k_F x) = \frac{v}{(2\pi)\rho} 2 \frac{\sin k_F x}{x}. \qquad (5.2.12)
$$

Therefore, we can write:

$$
\frac{\langle \Phi_{FS} | \hat{V} | \Phi_{FS} \rangle}{N} = \frac{1}{2} \rho \int_{-\infty}^{\infty} dx_{12} \, v(x_{12}) \, g(|x_{12}|). \qquad (5.2.13)
$$

The distribution function depends only on the distance $|x_{12}|$. Notice also that $l_{1D}(k_F x) = l_{1D}(-k_F x)$, which is why we can write $g(x_{12}) = g(|x_{12}|)$. The probability of having two particles at the same place is given by:

$$
g(0) = 1 - \frac{1}{v}, \qquad (5.2.14)
$$

because $l_{1D}^2(0) = 1$. The distribution functions go to $1 - O(1/x_{12}^2)$ when $x_{12} \to \infty$.

PROBLEM 5.3
Properties of the Slater function in 1D. Sequential condition of the two-body distribution function.

The two-body distribution function for the free Fermi sea in 1D with spin degeneracy $v = 2$ reads:

$$g(x_{12}) = 1 - \frac{l_{1D}^2(k_F|x_{12}|)}{v},$$ (5.3.1)

where

$$l_{1D}(k_F x) = \frac{v}{(2\pi)\rho} \int_{-k_F}^{k_F} dk \, e^{ikx} = \frac{v}{(2\pi)\rho} 2 \frac{\sin(k_F x)}{x}$$ (5.3.2)

is the Slater function in 1D and k_F is the Fermi momentum.

(a) Check that

$$\int_{-\infty}^{\infty} dx \, l_{1D}(k_F x) = \frac{v}{\rho}.$$ (5.3.3)

(b) Check that

$$\int_{-\infty}^{\infty} dx \, l_{1D}^2(k_F x) = \frac{v}{\rho}.$$ (5.3.4)

(c) Check that $g(x_{12})$ fulfills the sequential condition .

(a) We start with the normalization of the Slater function,

$$\int_{-\infty}^{\infty} dx \, l_{1D}(k_F x) = \int_{-\infty}^{\infty} dx \, \frac{v}{(2\pi)\rho} \int_{-k_F}^{k_F} e^{ikx} = \frac{v}{\rho} \int_{-k_F}^{k_F} dk \, \frac{1}{2\pi} \int_{-\infty}^{\infty} dx \, e^{ikx}. \quad (5.3.5)$$

One can identify the delta function $\delta(x)$ in the last expression,

$$\int_{-\infty}^{\infty} dx \, l_{1D}(k_F x) = \frac{v}{\rho} \int_{-k_F}^{k_F} dk \, \delta(k) = \frac{v}{\rho}. \quad (5.3.6)$$

(b) We start by substituting the definition of the Slater function,

$$\int_{-\infty}^{\infty} dx \, l_{1D}^2(k_F x) = \int_{-\infty}^{\infty} dx \, \frac{v}{(2\pi)\rho} \frac{v}{(2\pi)\rho} \left[\int_{-k_F}^{k_F} dk \, e^{ikx} \right] \left[\int_{-k_F}^{k_F} dk' \, e^{ik'x} \right]$$

$$= \int_{k_F}^{-k_F} dk \int_{-k_F}^{k_F} dk' \, \frac{v^2}{\rho^2(2\pi)} \frac{1}{2\pi} \int_{-\infty}^{\infty} dx \, e^{i(k+k')x}. \quad (5.3.7)$$

Now, one can identify the delta function $\delta(k+k')$ in the last expression. Then,

$$\int_{-\infty}^{\infty} dx \, l_{1D}^2(k_F x) = \frac{v^2}{\rho^2(2\pi)} \int_{-k_F}^{k_F} dk \int_{-k_F}^{k_F} dk' \, \delta(k+k') = \frac{v^2}{\rho^2(2\pi)} \int_{-k_F}^{k_F} dk$$

$$= \frac{v^2}{\rho^2(2\pi)} 2k_F = \frac{v}{\rho}, \tag{5.3.8}$$

where we have used Eq. (5.1.2) in the last equality. It is curious to notice that both integrals give the same result.

(c) Finally, let us calculate the sequential condition,

$$\rho \int_{-\infty}^{\infty} dx_{12} \, (g(x_{12}) - 1) = \rho \int_{-\infty}^{\infty} dx_{12} \left(1 - \frac{l_{1D}^2(k_F x_{12})}{v} - 1 \right)$$

$$= -\rho \int_{-\infty}^{\infty} dx_{12} \frac{l_{1D}^2(k_F x_{12})}{v} = -\rho \frac{v}{\rho v} = -1. \tag{5.3.9}$$

PROBLEM 5.4
Trapped fermions

N non-interacting fermions are trapped in a one-dimensional harmonic oscillator potential. The fermions are spin-polarized, with all spins oriented in a direction perpendicular to the confinement.

(a) Calculate the total, kinetic and harmonic potential energies.

(b) Calculate the chemical potential.

(c) Derive an energy density functional in terms of the local density. Assume that, locally, the system can be represented by a one-dimensional Fermi sea.

(d) Find the density profile by minimizing the energy functional with the condition that the number of particles is constant.

(e) Find the chemical potential associated with the Lagrange multiplier used to impose a fixed number of particles.

(f) Calculate the total energy, harmonic potential energy and kinetic energy from the density profile that minimizes the density functional and compare with the exact values.

(a) The single-particle spectrum of a one-dimensional harmonic oscillator is

$$\varepsilon_i = \left(\frac{1}{2} + i \right) \hbar\omega. \tag{5.4.1}$$

Then, the total energy of the ground state of the N fermions is obtained by summing the single-particle energies of the lowest filled single-particle levels:

$$\frac{E}{\hbar\omega} = \frac{1}{2} + \frac{3}{2} + \frac{5}{2} + \cdots + \left(\frac{1}{2} + N - 1\right)$$
$$= \frac{1}{2}(1 + 3 + 5 + \cdots + 2N - 1) = \frac{1}{2}N^2. \tag{5.4.2}$$

Therefore, using the virial theorem we can say that

$$E = \frac{1}{2}N^2\hbar\omega, \quad T = V_{HO} = \frac{1}{4}N^2\hbar\omega. \tag{5.4.3}$$

(b) The chemical potential at zero temperature is given as the difference of the ground-state energies of the system of N particles and the system of $N-1$ particles, which therefore coincides with the energy of the last occupied level, which in our case is $\mu = \varepsilon_N = (N - 1/2)\hbar\omega$. When the number of particles is large, the energy and the number of particles can be considered as continuous variables and the chemical potential can be approximately calculated as the derivative of the total energy with respect to the number of particles:

$$\mu = \frac{dE}{dN} = N\hbar\omega, \tag{5.4.4}$$

which is very similar to the exact value when N is large.

(c) To build the density functional, we should first recall which is the energy per unit length in a free Fermi gas in 1D. The relation between the linear density and the Fermi momentum is

$$N = \frac{L\nu}{2\pi}\int_{-k_F}^{k_F} dk = \frac{\nu L}{2\pi}2k_F \quad \rightarrow \rho = \frac{\nu k_F}{\pi}, \tag{5.4.5}$$

the energy is

$$E = \frac{L\nu}{2\pi}\int_{-k_F}^{k_F} dk \frac{\hbar^2 k^2}{2m} = N\frac{1}{3}\frac{\hbar^2 k_F^2}{2m}, \tag{5.4.6}$$

and the energy density for $\nu = 1$ is:

$$\frac{E}{L} = \frac{\hbar^2}{2m}\frac{\pi^2}{3}\rho^3. \tag{5.4.7}$$

Notice that we have assumed that, locally, the system can be seen as a Fermi sea (otherwise we could not use the above relation between ρ and k_F). Then, the energy functional can be written as:

$$E = \int dx \left(\frac{\hbar^2}{2m}\frac{\pi^2}{3}\rho^3 + \frac{1}{2}m\omega^2 x^2\rho\right). \tag{5.4.8}$$

In the next step we want to write the functional in units of the harmonic oscillator, i.e. the energy in units of $\hbar\omega$ and the length in units of the harmonic oscillator length: $b = (\hbar/(m\omega))^{1/2}$. To this end, we transform the density as $\bar{\rho}(x_1) = b\rho(x)$ where $x = bx_1$. Notice that this transformation keeps the density normalized to the number of particles:

$$\int dx_1 \bar{\rho}(x_1) = \int dx_1 b\rho(bx_1) = \int dy \frac{1}{b} b\,\rho(y) = N, \qquad (5.4.9)$$

where we have used the change $y = x_1 b$, and taken into account that the original density was normalized to the number of particles. Using the harmonic oscillator units, the functional can be written as

$$E_{HO} = \int dx_1 \left[\frac{\pi^2}{3} \frac{1}{2} \bar{\rho}(x_1)^3 + \frac{1}{2} x_1^2 \bar{\rho}(x_1) \right]. \qquad (5.4.10)$$

(d) Now, we should minimize the functional with respect to the density profile but keep the normalization of the density. To this end, we introduce a Lagrange multiplier, and require the total functional to be stationary with respect to small changes of $\bar{\rho}$:

$$\frac{\delta(E_{HO} - \mu \int dx_1 \bar{\rho}(x_1))}{\delta\bar{\rho}} = 0, \qquad (5.4.11)$$

to get the following condition:

$$\frac{1}{2}\pi^2 \bar{\rho}(x_1)^2 + \frac{1}{2} x_1^2 - \mu = 0. \qquad (5.4.12)$$

The density should be always positive or null, so:

$$\bar{\rho}(x_1) = \begin{cases} \frac{1}{\pi}\sqrt{2\mu - x_1^2} & |x_1| \le \sqrt{2\mu} \\ 0 & |x_1| > \sqrt{2\mu} \end{cases}.$$

(e) In the next step, we can determine the chemical potential by imposing that the integration of the density should provide the total number of particles:

$$\int_{-\sqrt{2\mu}}^{\sqrt{2\mu}} dx_1 \bar{\rho}(x_1) = \int_{-\sqrt{2\mu}}^{\sqrt{2\mu}} dx_1 \frac{1}{\pi}\sqrt{2\mu - x_1^2} = N. \qquad (5.4.13)$$

To perform this integral we should recall that

$$\int \sqrt{a + cx^2}\,dx = \frac{1}{2}x\sqrt{a + cx^2} + \frac{1}{2}\frac{a}{\sqrt{-c}} \arcsin\left(x\sqrt{-\frac{c}{a}}\right), \qquad (5.4.14)$$

and therefore

$$\int_{-\sqrt{2\mu}}^{\sqrt{2\mu}} dx_1 \bar{\rho}(x_1) = \frac{1}{\pi}\frac{1}{2}\left[x_1\sqrt{2\mu - x_1^2} + 2\mu \arcsin\left(\frac{x}{\sqrt{2\mu}}\right)\right]_{-\sqrt{2\mu}}^{\sqrt{2\mu}}$$

$$= \frac{1}{\pi}\mu\left[\frac{\pi}{2} - \left(-\frac{\pi}{2}\right)\right] = \mu = N.$$

Therefore, $\mu = N$ in harmonic oscillator units.

(f) Let us now proceed with the calculation of the harmonic potential energy:

$$\langle \hat{V}_{HO} \rangle = \int_{-\sqrt{2\mu}}^{\sqrt{2\mu}} dx_1 \frac{1}{2} x_1^2 \bar{\rho}(x_1) = \frac{1}{2} \int_{-\sqrt{2\mu}}^{\sqrt{2\mu}} dx_1 \, x_1^2 \frac{1}{\pi} \sqrt{2\mu - x_1^2}. \qquad (5.4.15)$$

Now, we should take into account the following integral:

$$\int dx x^2 \sqrt{a + cx^2} = \frac{1}{4} \frac{xa^3}{c} - \frac{1}{8} \frac{axu}{c} - \frac{1}{8} \frac{a^2}{c} I_1, \qquad (5.4.16)$$

where

$$u = \sqrt{a + cx^2} \quad , \quad I_1 = \frac{1}{\sqrt{-c}} \arcsin\left(x \sqrt{-\frac{c}{a}} \right) \quad , \quad c < 0. \qquad (5.4.17)$$

In our case, $a = 2\mu$ and $c = -1$, and we have:

$$\int_{-\sqrt{2\mu}}^{\sqrt{2\mu}} dx_1 \, x_1^2 \sqrt{2\mu - x_1^2} = -\frac{1}{4} x_1 \left(2\mu\right)^3 \Big|_{-\sqrt{2\mu}}^{\sqrt{2\mu}}$$

$$+ \frac{1}{8} 2\mu x_1 \sqrt{2\mu - x_1^2} \Big|_{-\sqrt{2\mu}}^{\sqrt{2\mu}}$$

$$+ \frac{1}{8} \left(2\mu\right)^2 \arcsin\left(\frac{x_1}{\sqrt{2\mu}} \right) \Big|_{-\sqrt{2\mu}}^{\sqrt{2\mu}} = \frac{\mu^2 2\pi}{4},$$

where the first two contributions to the integral are zero. Finally,

$$\langle \hat{V}_{HO} \rangle = \frac{1}{2} \frac{1}{\pi} \frac{\mu^2 2\pi}{4} = \frac{\mu^2}{4}, \qquad (5.4.18)$$

and taking into account that $\mu = N$, we have

$$\langle \hat{V}_{HO} \rangle = \frac{N^2}{4}. \qquad (5.4.19)$$

Now the kinetic energy is:

$$\langle \hat{T}_{HO} \rangle = \int_{-\sqrt{2\mu}}^{\sqrt{2\mu}} dx_1 \frac{\pi^2}{3} \frac{1}{2} \bar{\rho}(x_1)^3 = \frac{1}{\pi} \frac{1}{6} \int_{-\sqrt{2\mu}}^{\sqrt{2\mu}} dx_1 \left(2\mu - x_1^2\right)^{3/2}. \qquad (5.4.20)$$

One should take into account the following integral:

$$\int u^3 dx = \frac{1}{4} xu^3 + \frac{3}{8} axu + \frac{3}{8} a^2 I_1, \qquad (5.4.21)$$

where $u = \sqrt{a + cx^2}$, and therefore

$$\int_{-\sqrt{2\mu}}^{\sqrt{2\mu}} dx_1 \left(2\mu - x_1^2\right)^{3/2} = \frac{1}{4} x_1 \left(2\mu - x_1^2\right)^{3/2} \Big|_{-\sqrt{2\mu}}^{\sqrt{2\mu}} + \frac{3}{8} 2\mu x_1 \sqrt{2\mu - x_1^2} \Big|_{-\sqrt{2\mu}}^{\sqrt{2\mu}}$$

$$+ \frac{3}{8} \left(2\mu\right)^2 \arcsin\left(\frac{x}{\sqrt{2\mu}} \right) \Big|_{-\sqrt{2\mu}}^{\sqrt{2\mu}} = \frac{3}{2} \mu^2 \pi,$$

where the first two terms are identically zero. The kinetic energy is, finally:

$$\langle \hat{T}_{HO} \rangle = \frac{1}{6\pi} \frac{3}{2} \mu^2 \pi = \frac{\mu^2}{4} = \frac{N^2}{4}. \tag{5.4.22}$$

6 The Fermi sea in three dimensions at $T = 0$

The concept of the Fermi sea is central to understanding the electronic properties of many-body quantum systems. In one-dimensional systems, the Fermi sea is characterized by a simpler topology, with electrons occupying states along a linear momentum axis. In contrast, in three dimensions, the free Fermi sea takes on a more complex geometry, forming a spherical surface in momentum space. This difference profoundly influences the density of states, interactions between particles, and the emergent phenomena in these systems. Three-dimensional fermionic systems are governed by a rich spectrum of quantum states and exhibit different collective behaviors.

At zero temperature ($T = 0$), the Fermi sea is in its simplest and most well-defined form, with all states below the Fermi energy completely filled and those above completely empty. This limit case is particularly valuable for theoretical and computational studies, as it allows for analytic calculations to be performed with relative ease compared to finite-temperature scenarios. Despite its simplicity, the T=0 approximation remains useful, providing insights into the ground-state properties of quantum systems and serving as a foundation for understanding more complex finite-temperature properties.

This chapter includes four problems to explore in detail properties of the three-dimensional Fermi sea at zero temperature. The problems cover fundamental topics such as energy, thermodynamic properties and relativistic modifications of the thermodynamic properties. These problems provide a framework for exploring the theoretical and practical aspects of the system, offering essential tools for understanding both the classical and relativistic behavior of the Fermi sea.

PROBLEM 6.1
Energy and first derivatives.

Consider a uniform system of non-interacting non-relativistic fermions with number density ρ usually known as a free Fermi sea.

(a) Calculate the relation between the density and the Fermi momentum.

(b) Calculate the energy per particle as a function of the density.

(c) Calculate the pressure and the chemical potential.

(d) Calculate the incompressibility and the speed of sound.

(e) Make numerical applications for the case of nuclear matter at $\rho = 0.16$ fm^{-3} taking $\hbar^2/2m = 20.7344$ MeV fm^2.

(f) Do the same for liquid Helium-3 with $\rho = 0.0164$ Å^{-3} and $\hbar^2/2m = 8.03$ K Å^2.

(a) We assume that the fermions are enclosed in a cubic box of volume Ω and that the single-particle states are plane waves normalized to this volume, characterized by the momenta allowed by the periodic boundary conditions. The ground state is then given by a Slater determinant where the lowest single-particle states are filled up. Thus, we now start by calculating the Fermi momentum associated to the density. The total number of particles should be related to the total number of filled states,

$$N = \frac{\Omega}{(2\pi)^3} \nu \int_{|\vec{k}| \leq k_F} d\vec{k} = \frac{\Omega}{(2\pi)^3} \nu \frac{4}{3} \pi k_F^3, \qquad (6.1.1)$$

where ν is the spin/isospin degeneracy, i.e., $\nu = 4$ for nuclear matter, $\nu = 2$ for neutron matter, $\nu = 2$ for liquid ^{3}He, and $\nu = 2$ for electrons. Therefore

$$\rho = \frac{\nu k_F^3}{6\pi^2} \Rightarrow k_F = \left(\frac{6\pi^2 \rho}{\nu}\right)^{1/3} \qquad (6.1.2)$$

which establishes a relation between a macroscopic quantity, ρ, and a microscopic one, the Fermi momentum k_F.

(b) As the kinetic energy is a one-body operator, the expectation value in a Slater determinant is the sum of the expectation values of the kinetic energy operator in the occupied single-particle states,

$$e = \frac{E}{N} = \sum_{\vec{k}, m_s} \langle \vec{k}\, m_s | \frac{\hat{p}^2}{2m} | \vec{k}\, m_s \rangle \qquad (6.1.3)$$

where

$$\langle \vec{k} \, m_s | \frac{\hat{p}^2}{2m} | \vec{k} \, m_s \rangle = \langle \vec{k} | \frac{\hat{p}^2}{2m} | \vec{k} \rangle \, \langle m_s \mid m_s \rangle , \tag{6.1.4}$$

with

$$\langle \vec{k} | \frac{\hat{p}^2}{2m} | \vec{k} \rangle = \frac{1}{\Omega} \int_{\vec{r} \in \Omega} d\vec{r} \, e^{-i k \vec{r}} \left(-\frac{\hbar^2 \nabla^2}{2m} \right) e^{i k \vec{r}} = \frac{\hbar^2 k^2}{2m} . \tag{6.1.5}$$

Then,

$$e = \frac{\Omega}{(2\pi)^3} \frac{v}{N} \int_{|\vec{k}| \le k_F} d\vec{k} \, \frac{\hbar^2 k^2}{2m} = \frac{1}{N} \frac{\Omega}{(2\pi)^3} v \frac{\hbar^2}{2m} 4\pi \frac{k_F^5}{5} , \tag{6.1.6}$$

and thus writing N in terms of Ω and k_F results in,

$$e(\rho) = \frac{3}{5} \frac{\hbar^2 k_F^2}{2m} = \frac{3}{5} \frac{\hbar^2}{2m} \left(\frac{6\pi^2 \rho}{v} \right)^{2/3} . \tag{6.1.7}$$

(c) The first derivative of the total energy with respect to the volume, keeping the number of particles constant, provides the pressure. It is convenient to perform the derivative with respect to the density and to express the total energy as $E = N e(\rho)$,

$$P(\rho) = - \left(\frac{dE}{d\Omega} \right)_N = - \frac{de(\rho)}{d\rho} \left(\frac{d\rho}{d\Omega} \right)_N = \frac{\rho^2}{N} \frac{de(\rho)}{d\rho} . \tag{6.1.8}$$

Therefore,

$$P(\rho) = \frac{\hbar^2}{2m} \frac{2}{5} \left(\frac{6\pi^2}{v} \right)^{2/3} \frac{\rho^{5/3}}{N} . \tag{6.1.9}$$

On the other hand, the chemical potential

$$\mu(\rho) = \left(\frac{dE}{dN} \right) = e(\rho) + \frac{P(\rho)}{\rho} . \tag{6.1.10}$$

Therefore,

$$\mu(\rho) = \frac{\hbar^2}{2m} \frac{3}{5} \left(\frac{6\pi^2}{v} \right)^{2/3} \rho^{2/3} + \frac{\hbar^2}{2m} \frac{2}{5} \left(\frac{6\pi^2}{v} \right)^{2/3} \rho^{2/3} = \frac{\hbar^2 k_F^2}{2m} \tag{6.1.11}$$

i.e, the energy of the last occupied single-particle state.

(d) The compressibility modulus is the change in volume per unit volume when we vary the pressure,

$$\kappa = - \frac{1}{\Omega} \left(\frac{d\Omega}{dP} \right)_N . \tag{6.1.12}$$

Usually we calculate

$$\kappa^{-1} = - \left(\frac{dP}{d\Omega} \right)_N \Omega = \left(\frac{dP}{d\rho} \right) \rho = \frac{\hbar^2}{2m} \frac{2}{3} \left(\frac{6\pi^2}{v} \right)^{2/3} \frac{\rho^{5/3}}{N} . \tag{6.1.13}$$

Then, the speed of sound at $T = 0$ is related to the compressibility modulus by:

$$c_s^2 = \frac{\kappa^{-1}}{\rho m} = \frac{\hbar^2}{2m^2} \frac{2}{3} \left(\frac{6\pi^2}{v} \right)^{2/3} \rho^{2/3} = \frac{\hbar^2 k_F^2}{3m^2} = \frac{v_F^2}{3}. \qquad (6.1.14)$$

Finally,

$$c_s = \frac{\hbar\, k_F}{m\sqrt{3}} = \frac{v_F}{\sqrt{3}}. \qquad (6.1.15)$$

(e) Let us now consider nuclear matter at saturation density $\rho = 0.16$ fm^{-3}. Then $k_F = 1.33$ fm^{-1}. Taking $\hbar^2/2m = 20.7344$ MeV fm^2, we have $\langle \Phi_{FS}|T|\Phi_{FS}\rangle = 22.01$ MeV and the speed of sound can be expressed in terms of the speed of light c

$$\frac{c_s}{c} = \frac{\hbar c k_F}{\sqrt{3}mc^2} = 0.16 \qquad (6.1.16)$$

which is not relativistic.

(f) For the liquid ^{3}He, we have $\rho = 0.0164$ Å^{-3} and $\hbar^2/2m = 8.03$ K Å^2, which gives $k_F = 0.786$ Å^{-1}, the average kinetic energy, $\langle T \rangle = 2.976$ K and the speed of sound $c_s/c = 3.2\ 10^{-7}$.

PROBLEM 6.2
Symmetry energy of the free Fermi sea

Consider a uniform system of protons and neutrons, with total number density density $\rho = (Z+N)/\Omega$ where Z is the number of protons, N is the number of neutrons and Ω is the volume that encloses the system; and asymmetry $\Delta = (\rho_n - \rho_p)/\rho$ where ρ_n and ρ_p are the neutron and proton number densities, respectively. The system is modeled by an asymmetric Fermi sea. Take $\hbar^2/m = 41.49$ MeV fm^2.

(a) Calculate the total energy as a function of the number density and the asymmetry parameter Δ.

(b) Write the Taylor expansion of the energy for a given density in terms of the asymmetry parameter.

(c) Identify the contribution to the asymmetry energy. Give the result at the empirical saturation density of nuclear matter, $\rho_0 = 0.16$ fm^{-3}.

(d) Calculate the density derivative of the asymmetry energy at ρ_0.

(a) We proceed to work out the different questions. For the first one we start by calculating the Fermi momentum associated to the total density and the Fermi momenta associated to the proton and neutron components. To describe the system one can use the total density $\rho = \rho_N + \rho_Z$ and the asymmetry: $\Delta = (\rho_N - \rho_Z)/\rho$. At the same time, ρ_Z and ρ_n can be expressed in terms of the total density ρ and the asymmetry Δ:

$$\rho_Z = \frac{\rho}{2}(1 + \Delta), \quad \rho_N = \frac{\rho}{2}(1 - \Delta). \tag{6.2.1}$$

Then we have a global Fermi momentum associated to the total density:

$$k_F = \left(\frac{6\pi^2 \rho}{\nu} \right)^{1/3} = \left(\frac{6\pi^2 \rho}{4} \right)^{1/3}, \tag{6.2.2}$$

where we have considered the spin-isospin degeneracy, $\nu = 4$. In a similar manner, we define the Fermi momenta associated to protons and neutrons:

$$k_F^Z = \left(\frac{6\pi^2 \rho_Z}{\nu_s} \right)^{1/3} = \left(6\frac{\pi^2}{2} \frac{\rho}{2}(1 - \Delta) \right)^{1/3} = k_F(1 - \Delta)^{1/3}, \tag{6.2.3}$$

where $\nu_s = 2$ is the spin degeneracy, and

$$k_F^N = \left(\frac{6\pi^2 \rho_N}{\nu_s} \right)^{1/3} = \left(6\frac{\pi^2}{2} \frac{\rho}{2}(1 + \Delta) \right)^{1/3} = k_F(1 + \Delta)^{1/3}. \tag{6.2.4}$$

The total kinetic energy is the sum of the kinetic energies of protons and neutrons:

$$E = \frac{\Omega}{(2\pi)^3} \nu_s \int_{|\vec{k}| \leq k_F} d\vec{k} \, \frac{\hbar^2 k^2}{2m} + \frac{\Omega}{(2\pi)^3} \nu_s \int_{|\vec{k}| \leq k_F} d\vec{k} \, \frac{\hbar^2 k^2}{2m}, \tag{6.2.5}$$

where we have considered that the masses of protons and neutrons are the same, $m_z = m_N = m$. Performing the integrals:

$$
\begin{aligned}
E &= \frac{\Omega}{(2\pi)^3} 4\pi \, 2 \left[\frac{\hbar^2 (k_F^Z)^5}{10m} + \frac{\hbar^2 (k_F^N)^5}{10m} \right] \\
&= \frac{\Omega}{(2\pi)^3} 4\pi \frac{1}{10m} 2k_F^5 \left[(1 + \Delta)^{5/3} + (1 - \Delta)^{5/3} \right].
\end{aligned} \tag{6.2.6}
$$

As we are interested in the kinetic energy per particle,

$$
\begin{aligned}
e(k_F, \Delta) &= \frac{E}{N} = \frac{\frac{\Omega}{(2\pi)^3} 4\pi \frac{1}{10m} 2k_F^5 \left[(1 + \Delta)^{5/3} + (1 - \Delta)^{5/3} \right]}{\frac{\Omega}{(2\pi)^3} \frac{4}{3}\pi k_F^3 4} \\
&= \frac{3}{2} \frac{\hbar^2 k_F^2}{10m} \left[(1 + \Delta)^{5/3} + (1 - \Delta)^{5/3} \right] \\
&= \frac{3}{5} \frac{\hbar^2 k_F^2}{2m} \frac{1}{2} \left[(1 + \Delta)^{5/3} + (1 - \Delta)^{5/3} \right],
\end{aligned} \tag{6.2.7}
$$

where $N = \rho\Omega$. Notice that if $\Delta = 0$ we recover the kinetic energy per particle of the symmetric free Fermi sea.

(b) Now we perform the Taylor expansion of $\left[(1+\Delta)^{5/3} + (1-\Delta)^{5/3}\right]$ around $\Delta = 0$:

$$\left[(1+\Delta)^{5/3} + (1-\Delta)^{5/3}\right] \sim \left[1 + \frac{5}{3}\Delta + \frac{10}{18}\Delta^2 + \cdots + 1 - \frac{5}{3} + \frac{10}{18}\Delta^2 + \cdots\right]$$

$$\sim 2 + \frac{10}{9}\Delta^2 + \cdots \tag{6.2.8}$$

Notice that by symmetry, all terms with odd powers, in particular the linear term, in Δ disappear. Therefore, up to second order,

$$e(k_F, \Delta) = \frac{3}{5}\frac{\hbar^2 k_F^2}{2m}\left[1 + \frac{5}{9}\Delta^2 + \cdots\right]. \tag{6.2.9}$$

(c) By definition, the symmetry energy is associated with the coefficient of the term Δ^2 or equivalently, by looking at the Taylor expansion, with the one-half of the second derivative with respect to Δ of $e(\rho, \Delta)$:

$$e(\rho, \Delta) \sim \frac{3}{5}\frac{\hbar^2 k_F^2}{2m} + \frac{3}{5}\frac{\hbar^2 k_F^2}{2m}\frac{5}{9}\Delta^2 + \cdots, \tag{6.2.10}$$

and therefore

$$a_I = \frac{1}{3}\frac{\hbar^2 k_F^2}{2m}. \tag{6.2.11}$$

For the case of nuclear matter, with $k_F = 1.36\ \text{fm}^{-1}$ which corresponds to the nuclear saturation density, and taking $\hbar^2/m = 41.46\ \text{MeV fm}^2$, we have that $a_I = 12.78\ \text{MeV}$. Notice that a commonly accepted value for nuclear matter is 30 MeV, which means that correlations will have an important role in the evaluation of this quantity and that the free Fermi gas model is rather limited in this case. If only the second power of Δ would be present in the expansion, it would be possible to calculate the symmetry energy as $a_I \sim e(\rho, \Delta = 1) - e(\rho, \Delta = 0)$. It is therefore interesting to evaluate the difference between the two procedures:

$$\tilde{a}_I = e(\rho, \Delta = 1) - e(\rho, \Delta = 0) = \frac{3}{5}\frac{\hbar^2 k_F}{2m}(2^{2/3} - 1) \tag{6.2.12}$$

and the relative error :

$$\frac{a_I - \tilde{a}_I}{a_I} = \frac{\frac{5}{9} - (2^{2/3} - 1)}{\frac{5}{9}} = -0.06. \tag{6.2.13}$$

(d) Also of interest is to provide the behavior of the density derivative of a_I. To this end, it is useful to express the symmetry energy in terms of the density instead than the Fermi momentum:

$$e(\rho, \Delta) = \frac{3}{5}\frac{\hbar^2}{2m}\left(\frac{6\pi^2\rho}{\nu}\right)^{2/3} + \frac{3}{5}\frac{\hbar^2}{2m}\left(\frac{6\pi^2\rho}{\nu}\right)^{2/3}\frac{5}{9}\Delta^2 + \cdots \tag{6.2.14}$$

Therefore,

$$a_I = \frac{1}{3} \frac{\hbar^2}{2m} \left(\frac{6\pi^2 \rho}{v} \right)^{2/3} \frac{5}{9} \tag{6.2.15}$$

and

$$\frac{da_I(\rho)}{\rho} = \frac{1}{3} \frac{\hbar^2}{2m} \left(\frac{6\pi^2}{v} \right)^{2/3} \frac{2}{3} \rho^{-1/3} \frac{5}{9} = \frac{2}{3} \frac{a_I}{\rho^{1/3}}. \tag{6.2.16}$$

The value at $\rho = 0.16 \text{ fm}^{-3}$,

$$\frac{da_I}{d\rho} = 15.70 \text{ MeV fm}^3 \tag{6.2.17}$$

PROBLEM 6.3
The energy of the free Fermi sea as a lower bound of the kinetic energy

Consider a non-relativistic uniform system of fermions characterized by a density ρ. Show that the energy of the free Fermi sea associated to that system is a lower bound to the expectation value of the kinetic energy of the system.

The kinetic energy operator

$$\hat{T} = -\sum_i \frac{\hbar^2 \nabla_i^2}{2m} \tag{6.3.1}$$

is the Hamiltonian, $\hat{H}_0 \equiv \hat{T}$, of the free Fermi sea and its ground state, ϕ_0 is the Slater determinant built with plane waves filling up all momenta up to the Fermi level. The energy per particle of the ground state is the well known:

$$\hat{H}_0 |\phi_0\rangle = E_0 |\phi_o\rangle, \quad E_0 = \frac{3}{5} \frac{\hbar^2 k_F^2}{2m} = \frac{3}{5} \frac{\hbar^2}{2m} \left(\frac{6\pi^2 \rho}{v} \right)^{2/3} \tag{6.3.2}$$

If we consider now the ground state $|\Psi_0\rangle$ of another Hamiltonian, $\hat{H} = \hat{T} + \hat{V}$, and calculate the expectation value of $\hat{H}_0 \equiv \hat{T}$ in this wavefunction, the variational principle tells us that

$$\langle \Psi_0 | \hat{T} | \Psi_0 \rangle = \langle \Psi_0 | \hat{H}_0 | \Psi_0 \rangle = E_{var} \geq E_0 \tag{6.3.3}$$

and therefore, E_0 is a lower bound of the expectation value of the kinetic energy in the ground state of $\hat{H}$.

PROBLEM 6.4
Thermodynamic properties of the relativistic three-dimensional free Fermi sea.

Consider a uniform system of non-interacting spin-1/2 fermions in three dimensions at zero temperature. Use the relativistic dispersion relation between momentum and energy:

$$\varepsilon(k) = \sqrt{m^2c^4 + \hbar^2k^2c^2} = mc^2\sqrt{1+x^2}, \qquad (6.4.1)$$

where
$$x = \frac{\hbar c k}{mc^2}. \qquad (6.4.2)$$

This parameter measures the relativistic effects. For $x \ll 1$, relativistic effects are small.

(a) Calculate the energy density.

(b) Calculate the chemical potential.

(c) Calculate the pressure.

(a) To calculate the energy density, one should sum up all the single-particle energies of the occupied states. To this end,

$$\varepsilon_\Omega = \frac{E}{\Omega} = V\frac{1}{(2\pi)^3}\int d\vec{k}\,\Theta(k_F - k)\,\varepsilon(k) \qquad (6.4.3)$$

Now, using the dispersion relation we can write:

$$\begin{aligned}
\varepsilon_\Omega &= V\frac{1}{(2\pi)^3}mc^2\int d\vec{k}\,\Theta(k_F - k)\sqrt{1+x(k)^2}\\
&= V\frac{1}{(2\pi)^3}mc^2 4\pi\int_{k\leq k_F} dk\,k^2\sqrt{1+x(k)^2}\\
&= V\frac{1}{(2\pi)^3}\frac{m^2c^4}{\hbar^2c^2}\frac{mc^2}{\hbar c}mc^2 4\pi\int_0^{x_F} x^2\sqrt{1+x^2}\,dx\\
&= \frac{(mc^2)^4}{(\hbar c)^3}V\frac{1}{2\pi^2}\int_0^{x_F} x^2\sqrt{1+x^2}\,dx = \varepsilon_0\,\bar{\varepsilon}(x_F), \qquad (6.4.4)
\end{aligned}$$

where $\varepsilon_0 = (mc^2)^4/(\hbar c)^3$, and

$$
\begin{aligned}
\bar{\varepsilon}(x_F) &= \frac{v}{2\pi^2} \int_0^{x_F} x^2 \sqrt{1+x^2}\, dx \\
&= \frac{v}{16\pi^2} \left[x_F(1 + 2x_F^2)\sqrt{1 + x_F^2} - \ln\left(x_F + \sqrt{x_F^2 + 1}\right) \right] \\
&= \frac{v}{16\pi^2} \left[x_F(1 + 2x_F^2)\sqrt{1 + x_F^2} - \sinh^{-1}(x_F) \right].
\end{aligned}
\tag{6.4.5}
$$

The non-relativistic limit is recovered for $x_F \ll 1$ which should coincide with

$$
\frac{E}{\Omega} = \rho mc^2 + \frac{3}{5}\frac{\hbar^2 k_F^2}{2m}\rho = \frac{v}{6\pi^2}mc^2 k_F^3 + \frac{3}{5}\frac{\hbar^2}{2m}\frac{v}{6\pi^2}k_F^5,
\tag{6.4.6}
$$

In fact, doing the series expansion of $\bar{\varepsilon}(x_F)$ around $x_F = 0$ one gets:

$$
\bar{\varepsilon}(x_F) \sim \frac{v}{16\pi^2}\left(\frac{8}{3}x_F^3 + \frac{4}{5}x_F^5 + \cdots \right),
\tag{6.4.7}
$$

which can be expressed as

$$
\begin{aligned}
\varepsilon_\Omega &= \varepsilon_0\, \bar{\varepsilon}(x_F) \sim \frac{(mc^2)^4}{(\hbar c)^3}\frac{v}{16\pi^2}\left[\frac{8}{3}\frac{(\hbar c)^3}{(mc^2)^3}k_F^3 + \frac{4}{5}\frac{(\hbar c)^5}{(mc^2)^5}k_F^5 + \cdots \right] \\
&= mc^2\frac{v}{6\pi^2}k_F^3 + \frac{\hbar^2 k_F^2}{2m}\frac{3}{5}\frac{v}{6\pi^2}k_F^3 = mc^2\rho + \frac{3}{5}\frac{\hbar^2 k_F^2}{2m}\rho.
\end{aligned}
\tag{6.4.8}
$$

(b) The chemical potential is given by

$$
\mu = \left(\frac{\partial E}{\partial N}\right)_\Omega = \left(\frac{\partial(\varepsilon_\Omega \Omega)}{\partial N}\right)_\Omega = \Omega\frac{\partial \varepsilon_\Omega(\rho)}{\partial\rho}\frac{\partial\rho}{\partial N} = \frac{\partial \varepsilon_\Omega(\rho)}{\partial\rho},
\tag{6.4.9}
$$

which in terms of x_F is expressed as

$$
\mu = \varepsilon_0\frac{\partial\bar{\varepsilon}(x_F)}{\partial x_F}\frac{\partial x_F}{\partial\rho}
\tag{6.4.10}
$$

where

$$
\rho = \frac{v\, k_F^3}{6\pi^2}, \quad k_F = \left(\frac{6\pi^2\rho}{v}\right)^{1/3}, \quad \frac{\partial x_F}{\partial\rho} = \frac{\hbar c}{mc^2}\frac{1}{3}\frac{6\pi^2}{v}\frac{1}{k_F^2}.
\tag{6.4.11}
$$

Then,

$$\frac{\partial \bar{\varepsilon}(x_F)}{\partial x_F} = \frac{v}{16\pi^2} \frac{\partial}{\partial x_F} \left[(x_F + 2x_F^3)\sqrt{1 + x_F^2} - \ln\left(x_F + \sqrt{1 + x_F^2}\right) \right]$$

$$= \frac{v}{16\pi^2} \left[(1 + 6x_F)\sqrt{1 + x_F^2} + (x_F + 2x_F^3)\frac{2x_F}{2\sqrt{1 + x_F^2}} \right. \tag{6.4.12}$$

$$\left. - \frac{1}{x_F + \sqrt{1 + x_F^2}} \left(1 + \frac{2x_F}{2\sqrt{1 + x_F^2}} \right) \right]$$

$$= \frac{v}{2\pi^2} x_F^2 \sqrt{1 + x_F^2}. \tag{6.4.13}$$

Then,

$$\mu = \varepsilon_0 \frac{\partial \bar{\varepsilon}(x_F)}{\partial x_F} \frac{\partial x_F}{\partial \rho} = \frac{(mc^2)^4}{(\hbar c)^3} \frac{v}{2\pi^2} x_F^2 \sqrt{1 + x_F^2} \frac{\hbar c}{mc^2} \frac{1}{3} \frac{6\pi^2}{v} \frac{1}{k_F^2}$$

$$= mc^2 \sqrt{1 + x_F^2} = \varepsilon(x_F), \tag{6.4.14}$$

which is nothing other than the energy of the highest occupied level. The non-relativistic limit is obtained when $x_F \ll 1$, then

$$\mu(x_F \ll 1) = \lim_{x_F \to 0} mc^2 \sqrt{1 + x_F^2} \sim mc^2\left(1 + \frac{x_F^2}{2}\right) = mc^2 + \frac{\hbar^2 k_F^2}{2m} \tag{6.4.15}$$

(c) The pressure is given by

$$P = -\left(\frac{\partial E}{\partial \Omega}\right)_N = -\left(\frac{\partial(\Omega \varepsilon_\Omega(\rho))}{\partial \Omega}\right)_N =$$

$$-\varepsilon_\Omega(\rho) - \Omega \frac{\partial \varepsilon_\Omega}{\partial \rho} \left(\frac{\partial \rho}{\partial \Omega}\right)_N = -\varepsilon_\Omega + \rho \frac{\partial \varepsilon_\Omega}{\partial \rho} \tag{6.4.16}$$

which can be expressed as

$$P = -\varepsilon_\Omega + \mu\rho. \tag{6.4.17}$$

Now, taking into account that, $\varepsilon_\Omega(\rho) = \varepsilon_0 \bar{\varepsilon}(x_F)$, we can write

$$\rho\mu = \rho \frac{\partial \varepsilon_\Omega}{\partial \rho} = \rho \, \varepsilon_0 \frac{\partial \bar{\varepsilon}(x_F)}{\partial x_F} \frac{\partial x_F}{\partial \rho}$$

$$= \frac{v}{(2\pi)^3} \frac{4}{3} \pi k_F^3 \, \varepsilon_0 \frac{\partial \bar{\varepsilon}(x_F)}{\partial x_F} \frac{\hbar c}{mc^2} \frac{1}{3} \frac{6\pi^2}{v} \frac{1}{k_F^2}$$

$$= k_F \frac{\hbar c}{mc^2} \frac{1}{3} \varepsilon_0 \frac{\partial \bar{\varepsilon}(x_F)}{\partial x_F} = x_F \frac{1}{3} \varepsilon_0 \frac{\partial \bar{\varepsilon}(x_F)}{\partial x_F}. \tag{6.4.18}$$

Then,

$$P = P_0 \bar{P}(x_F) \tag{6.4.19}$$

with

$$\bar{P}(x_F) = \frac{1}{3} x_F \bar{\varepsilon}'(x_F) - \bar{\varepsilon}(x_F) \tag{6.4.20}$$

and $P_0 = \varepsilon_0$, and finally

$$\bar{\varepsilon}'(x_F) = \frac{\partial \bar{\varepsilon}(x_F)}{\partial x_F} = \frac{v}{2\pi^2} x_F^2 \sqrt{1 + x_F^2}. \tag{6.4.21}$$

7 Free Fermi sea at finite temperature

At finite temperatures, the sharp distinction between occupied and unoccupied states near the Fermi surface becomes blurred, as the filling of the states is ruled by the Fermi distribution . This leads to significant changes in the physical properties of the system. Analyzing these effects is crucial for understanding phenomena like electrical and thermal conductivity, as well as the response of materials to external perturbations. Moreover, working at nonzero temperatures allows us to study transitions between quantum and classical regimes, shedding light on the statistical mechanics that govern systems from condensed matter to nuclear physics and astrophysics.

This chapter includes problems that explore the effects of temperature on the Fermi sea, bridging the gap between the zero-temperature idealization and realistic finite-temperature scenarios. Important tools are included to compute thermal averages, some thermodynamic properties, and derive temperature-dependent corrections in the high and low temperature limits, in systems where thermal excitations play a crucial role in shaping their behavior.

PROBLEM 7.1
Specific heat at the low temperature limit of the free Fermi sea.

At low temperature, i.e. in the degenerate limit, the specific heat per particle at constant volume Ω of the free Fermi sea can be expressed as

$$C_\Omega(\rho,T) = \frac{\pi^2 T}{2\varepsilon_F}, \qquad (7.1.1)$$

where $\varepsilon_F = \hbar^2 k_F^2/2m$. Use this expression to calculate the low temperature behavior of:

(a) The energy per particle of the free Fermi sea.

(b) The entropy per particle.

(c) The free energy per particle.

(d) The pressure.

(e) The chemical potential.

(a) The increment of energy between two temperatures at a given density can be expressed as

$$e(\rho,T) = e(\rho,T=0) + \int_0^T C_\Omega(\rho,T')dT', \qquad (7.1.2)$$

which in our case is:

$$e(\rho,T) = \frac{3}{5}\varepsilon_F + \int_0^T \frac{\pi^2}{2\varepsilon_F}T'dT' = \frac{3}{5}\varepsilon_F + \pi^2\varepsilon_f\frac{1}{4}\left(\frac{T}{\varepsilon_F}\right)^2, \qquad (7.1.3)$$

where $\varepsilon_F = \hbar^2 k_F^2/2m$ is the Fermi energy of the free Fermi sea at $T = 0$. Therefore, the energy increases quadratically with temperature as

$$e(\rho,T) = \frac{3}{5}\varepsilon_F\left(1 + \pi^2\frac{5}{12}\left(\frac{T}{\varepsilon_F}\right)^2\right). \qquad (7.1.4)$$

(b) By definition,

$$s = \int_0^T \frac{C_\Omega(\rho,T')}{T'}dT' = \int_0^T \frac{\pi^2 T'}{2\varepsilon_F}\frac{1}{T'}dT'. \qquad (7.1.5)$$

Performing the integral, we get:

$$s = \frac{\pi^2}{2\varepsilon_f}\int_0^T dT' = \frac{\pi^2}{2\varepsilon_F}T, \qquad (7.1.6)$$

which is linear in T.

(c) The free energy per particle is defined as $f(\rho,T) = e(\rho,T) - Ts(\rho,T)$, which in our case is

$$f(\rho,T) = \frac{3}{5}\varepsilon_F + \varepsilon_F \frac{\pi^2}{4}\left(\frac{T}{\varepsilon_F}\right)^2 - \frac{\pi^2}{2\varepsilon_F}T^2 = \frac{3}{5}\varepsilon_F - \frac{1}{\varepsilon_F}\frac{\pi^2}{4}T^2, \qquad (7.1.7)$$

which can be written as

$$f(\rho,T) = \frac{3}{5}\varepsilon_F\left(1 - \frac{5\pi^2}{12}\left(\frac{T}{\varepsilon_F}\right)^2\right). \qquad (7.1.8)$$

At very low temperatures, the free energy decreases in respect to $T = 0$ due to the fact that the increase of entropy dominates over the increase of energy.

(d) The pressure is related to the free energy by

$$P(\rho,T) = \rho^2\frac{\partial f(\rho,T)}{\partial\rho}. \qquad (7.1.9)$$

To perform this derivative it is useful to have ε_F expressed in terms of the density, which by Eq. (6.1.2) can be put as

$$\varepsilon_F = \frac{\hbar^2}{2m}(6\pi^2 v^{-1})^{2/3}\rho^{2/3}, \qquad (7.1.10)$$

where v is the spin-isospin degeneracy. Then,

$$\begin{aligned}
P(\rho,T) &= \rho^2\frac{\partial}{\partial\rho}\left(\frac{3}{5}\varepsilon_F\left(1 - \frac{5\pi^2}{12}\left(\frac{T}{\varepsilon_F}\right)^2\right)\right) \\
&= \rho^2\frac{3}{5}\frac{\hbar^2}{2m}(6\pi^2 v^{-1})^{2/3}\frac{2}{3}\rho^{-1/3}\left(1 - \frac{5\pi^2}{12}\left(\frac{T}{\varepsilon_F}\right)^2\right) \\
&\quad + \rho^2\frac{3}{5}\varepsilon_F\frac{5\pi^2}{12}T^2\frac{2}{\varepsilon_F^3}\frac{\hbar^2}{2m}(6\pi^2 v^{-1})^{2/3}\frac{2}{3}\rho^{-1/3} \\
&= \frac{2}{5}\rho\varepsilon_F + \rho\frac{T^2}{\varepsilon_F}\pi^2\left(-\frac{2}{12} + \frac{4}{12}\right),
\end{aligned}$$

which finally is expressed as

$$P(\rho,T) = P(\rho,T = 0) + \rho\frac{\pi^2}{6}\frac{T^2}{\varepsilon_F}, \qquad (7.1.11)$$

where we have put in evidence the value of the pressure at $T = 0$,

$$P(\rho,T = 0) = \frac{2}{5}\rho\varepsilon_F. \qquad (7.1.12)$$

Finally,

$$P(\rho,T) = P(\rho,T=0)\left(1+\frac{5\pi^2}{12}\left(\frac{T}{\varepsilon_F}\right)^2\right). \qquad (7.1.13)$$

The pressure increases quadratically with the temperature.

(e) The chemical potential is related to the free energy and the pressure by

$$\mu(\rho,T) = f(\rho,T) + \frac{P(\rho,T)}{\rho}. \qquad (7.1.14)$$

Taking into account the expressions of the free-energy and the pressure we get

$$\mu(\rho,T) = \frac{3}{5}\varepsilon_F\left(1-\frac{5\pi^2}{12}\left(\frac{T}{\varepsilon_F}\right)^2\right) + \frac{2}{5}\varepsilon_F\left(1+\frac{5\pi^2}{12}\left(\frac{T}{\varepsilon_F}\right)^2\right), \qquad (7.1.15)$$

which finally can be written as

$$\mu(\rho,T) = \varepsilon_F - \pi^2\varepsilon_F\left(\frac{T}{\varepsilon_F}\right)^2\left(\frac{1}{4}-\frac{1}{6}\right) = \varepsilon_F\left(1-\frac{\pi^2}{12}\left(\frac{T}{\varepsilon_F}\right)^2\right). \qquad (7.1.16)$$

The chemical potential decreases quadratically with temperature in respect to the value at $T=0$ because the decrease of the free energy dominates over the increase of the pressure term.

PROBLEM 7.2
Free Fermi sea at finite temperature

Consider a uniform free Fermi sea of non-interacting, non-relativistic fermions of density ρ at temperature T. The occupation of a given momentum $\vec{k}$ is given by

$$n(\vec{k}) = \frac{\nu}{1+e^{\frac{\varepsilon(k)-\mu}{T}}}, \qquad (7.2.1)$$

where $\varepsilon(k) = \hbar^2k^2/2m$, ν is the spin-isospin degeneracy and μ is the chemical potential.

(a) Give an expression of the density in terms of the Fermi integrals defined as

$$J_\nu(\eta) = \int_0^\infty \frac{x^\nu dx}{1+e^{x-\eta}}. \qquad (7.2.2)$$

(b) Obtain the energy per particle in terms of the Fermi integrals.

(a) The total number of particles can be expressed as the sum of all occupations,

$$N = \sum_{\vec{k}} \frac{v}{1 + e^{\frac{\varepsilon(k)-\mu}{T}}} \, . \tag{7.2.3}$$

To perform the summation it is useful to transform the sum into an integral by means of

$$\sum_{\vec{k}} \longrightarrow \frac{\Omega}{(2\pi)^3} \int d\vec{k} \, , \tag{7.2.4}$$

where Ω is the volume of the enclosing box, or equivalently,

$$\sum_{\vec{p}} \longrightarrow \frac{\Omega}{(2\pi\hbar)^3} \int d\vec{p} \, . \tag{7.2.5}$$

Then,

$$N = \frac{\Omega}{(2\pi\hbar)^3} \int d\vec{p} \frac{v}{1 + e^{\frac{\varepsilon(p)-\mu}{T}}} \, . \tag{7.2.6}$$

Therefore,

$$\frac{dN}{dp} = \frac{4\pi}{(2\pi\hbar)^3} \frac{v\Omega p^2}{1 + e^{\frac{\varepsilon(p)-\mu}{T}}} \, , \tag{7.2.7}$$

where $\varepsilon(p) = p^2/2m$, and $dp = (2m)^{1/2} d\varepsilon/(2\varepsilon^{1/2})$. Then the density of states per unit energy can be expressed as

$$\frac{dN}{d\varepsilon} = \frac{4\pi v\Omega}{(2\pi\hbar)^3} \frac{2m\varepsilon(2m)^{1/2}}{1 + e^{\frac{\varepsilon-\mu}{T}}} \frac{1}{2\varepsilon^{1/2}} \, , \tag{7.2.8}$$

and the total number of particles:

$$N = \int_0^\infty dN_\varepsilon = \frac{v\Omega m^{3/2}}{2^{1/2}\pi^2\hbar^3} \int_0^\infty d\varepsilon \frac{\varepsilon^{1/2}}{1 + e^{\frac{\varepsilon-\mu}{T}}} \, . \tag{7.2.9}$$

Now, it is convenient to do the following change of variables: $z = \varepsilon/T$, and $d\varepsilon = dzT$ to get

$$N = \frac{v\Omega m^{3/2}T^{3/2}}{2^{1/2}\pi^2\hbar^3} \int_0^\infty \frac{z^{1/2}dz}{1 + e^{(z-\mu/T)}} \, . \tag{7.2.10}$$

Finally,

$$\rho = \frac{v}{2^{1/2}\pi^2} \left(\frac{mT}{\hbar^2}\right)^{3/2} \int_0^\infty \frac{z^{1/2}}{1 + e^{z-\eta}} \, , \tag{7.2.11}$$

with $\eta = \mu/T$, which can be expressed in terms of the Fermi integrals as

$$\rho = \frac{v}{2^{1/2}\pi^2} \left(\frac{mT}{\hbar^2}\right)^{3/2} J_{1/2}(\eta) \, . \tag{7.2.12}$$

(b) For the total energy we must sum the energies of each state times their occupations:

$$E = \int_0^\infty \varepsilon \, dN_\varepsilon = \frac{v\Omega m^{3/2}}{2^{1/2}\pi^2\hbar^3} \int_0^\infty \frac{\varepsilon^{3/2}}{1+e^{(\varepsilon-\mu)/T}} d\varepsilon. \tag{7.2.13}$$

Performing the same change of variable as before,

$$E(\eta) = \frac{v\Omega m^{3/2}T^{5/2}}{2^{1/2}\pi^2\hbar^3} \int_0^\infty \frac{z^{3/2}}{1+e^{z-\eta}} dz. \tag{7.2.14}$$

The energy density will be:

$$e_\Omega = \frac{E(\eta)}{\Omega} = \frac{vm^{3/2}T^{5/2}}{2^{1/2}\pi^2\hbar^3} \int_0^\infty \frac{z^{3/2}}{1+e^{z-\eta}} dz = \frac{vm^{3/2}T^{5/2}}{2^{1/2}\pi^2\hbar^3} J_{3/2}(\eta), \tag{7.2.15}$$

and the energy per particle,

$$e = \frac{E(\eta)}{N} = \frac{1}{\rho} \frac{vm^{3/2}T^{5/2}}{2^{1/2}\pi^2\hbar^3} J_{3/2}(\eta). \tag{7.2.16}$$

Usually we give ρ and T; then we determine η by inverting the relation between ρ and η, which permits us to obtain the chemical potential, $\mu = \eta T$, and afterwards we calculate the energy per particle $e(\rho, T)$.

PROBLEM 7.3
High T limit of the free Fermi sea

Consider a uniform free Fermi sea of non-interacting, non-relativistic fermions of density ρ at temperature T. For a given density, when T becomes very large or for a given temperature, when ρ becomes very small, then $z = e^{\mu/T}$ is such that $0 < z \ll 1$. In this limit, the function

$$f_n(z) = \frac{1}{\Gamma(n)} \int_0^\infty \frac{x^{n-1}}{1+e^x z^{-1}} dx \tag{7.3.1}$$

can be expanded as

$$f_n(z) = \frac{1}{\Gamma(n)} \int_0^\infty x^{n-1} \sum_{l=1}^\infty (-1)^{l-1}(ze^{-x})^l dx = z - \frac{z^2}{2^n} + \frac{z^3}{3^n} + \dots \tag{7.3.2}$$

(a) Using the first term of this expansion find the chemical potential from the expression

$$\rho = \frac{v}{2^{1/2}\pi^2} \left(\frac{mT}{\hbar^2}\right)^{3/2} J_{1/2}(\eta), \tag{7.3.3}$$

where $\eta = \mu/T$, and

$$J_{1/2}(\eta) = \int_0^\infty \frac{x^{1/2}}{1+e^x e^{-\eta}} dx. \tag{7.3.4}$$

(b) Calculate the energy per particle in this limit.

(c) Calculate the pressure in this limit.

(a) The density and the chemical potential at a given temperature are related by

$$\rho = \frac{v}{2^{1/2}\pi^2}\left(\frac{mT}{\hbar^2}\right)^{3/2} J_{1/2}(\eta),$$ (7.3.5)

where $\eta = \mu/T$. Using the expansion of $f_{3/2}(z)$ we can write,

$$\rho = \frac{v}{2^{1/2}\pi^2}\left(\frac{mT}{\hbar^2}\right)^{3/2}\int_0^\infty \frac{x^{3/2-1}}{1+e^x e^{-\eta}}\,dx \approx \frac{v}{2^{1/2}\pi^2}\left(\frac{mT}{\hbar^2}\right)^{3/2}\Gamma(3/2)z,$$ (7.3.6)

where $z = e^\eta$ and $\Gamma(3/2) = \pi^{1/2}/2$. Therefore,

$$\eta = \ln\left[\frac{\rho}{v}\left(\frac{2\pi\hbar^2}{mT}\right)^{3/2}\right].$$ (7.3.7)

Taking into account that $\eta = \mu/T$, we have

$$\mu = T\ln\left[\frac{\rho}{v}\left(\frac{2\pi\hbar^2}{mT}\right)^{3/2}\right],$$ (7.3.8)

which coincides with the chemical potential of a classical ideal gas. Notice that $\mu \to -\infty$ when $\rho \to 0$ for a given T, or when $T \to \infty$ for a given ρ.

(b) The energy per particle can be expressed as:

$$e = \frac{1}{\rho}\frac{vm^{3/2}T^{5/2}}{2^{1/2}\pi^2\hbar^3}\int_0^\infty \frac{x^{3/2}}{1+e^x e^{-\eta}}\,dx.$$ (7.3.9)

Performing the expansion of $J_{3/2}(z)$, when $z = e^\eta \to 0$,

$$J_{3/2}(z) = \int_0^\infty \frac{x^{3/2}}{1+e^x z^{-1}} = \Gamma(5/2)\frac{1}{\Gamma(5/2)}\int_0^\infty \frac{x^{5/2-1}}{1+e^x z^{-1}}\,dx \approx \Gamma(5/2)z$$
$$= \Gamma(5/2)e^\eta.$$

Then,

$$e = \frac{1}{\rho}\frac{vm^{3/2}T^{5/2}}{2^{1/2}\pi^2\hbar^3}\Gamma(5/2)e^\eta.$$ (7.3.10)

Taking into account that in this limit,

$$\eta = \ln\left[\frac{\rho}{v}\left(\frac{2\pi\hbar^2}{mT}\right)^{3/2}\right],$$

one gets,

$$e \sim \frac{1}{\rho}\frac{vm^{3/2}T^{5/2}}{2^{1/2}\pi^2\hbar^3}\Gamma(5/2)\frac{\rho}{v}\left(\frac{2\pi\hbar^2}{mT}\right)^{3/2}, \qquad (7.3.11)$$

with $\Gamma(5/2) = \pi^{1/2}3/4$, and therefore

$$e \sim \frac{3}{2}T, \qquad (7.3.12)$$

which is the energy per particle of an ideal classical gas, which depends linearly on temperature. Notice that the expression is in agreement with the equipartition theorem that assigns $1/2T$ for each degree of freedom and therefore in 3D one gets $3T/2$.

(c) In the grand canonical ensemble,

$$\frac{P\Omega}{T} = \sum_\varepsilon \ln(1 + ze^{-\beta\varepsilon}), \qquad (7.3.13)$$

where $z = e^{\mu/T}$ and $\beta = 1/T$. To perform the summation over all the energy states it is convenient to transform the summation into an integral:

$$\sum_\varepsilon \to \frac{\Omega v 4\pi}{(2\pi\hbar)^3}(2m)^{3/2}\frac{1}{2}\int_0^\infty \varepsilon^{1/2}d\varepsilon. \qquad (7.3.14)$$

Therefore

$$\sum_\varepsilon \ln(1 + ze^{-\beta\varepsilon}) = \frac{\Omega v 4\pi}{(2\pi\hbar)^3}(2m)^{3/2}\frac{1}{2}\int_0^\infty \varepsilon^{1/2}\ln(1 + ze^{-\beta\varepsilon})d\varepsilon. \qquad (7.3.15)$$

Performing the change of variables $x = \varepsilon/T$, $d\varepsilon = Tdx$ we have:

$$\sum_\varepsilon \ln(1 + ze^{-\beta\varepsilon}) = \frac{\Omega v 4\pi}{(2\pi\hbar)^3}(2mT)^{3/2}\frac{1}{2}\int_0^\infty \ln(1 + ze^{-x})x^{1/2}dx. \qquad (7.3.16)$$

Integrating by parts, with $ds = x^{1/2}dx$ then $s = 2/3x^{3/2}$ and $u = \ln(1 + ze^{-x})$ which implies $du = -ze^{-x}/(1 + ze^{-x}) = -1/(1 + z^{-1}e^x)$ we get

$$\sum_\varepsilon \ln(1 + ze^{-\beta\varepsilon}) = \frac{\Omega v 4\pi}{(2\pi\hbar)^3}(2mT)^{3/2}\frac{1}{2}\frac{2}{3}\int_0^\infty \frac{x^{3/2}}{1 + z^{-1}e^x}dx, \qquad (7.3.17)$$

which can be written as:

$$\frac{\Omega v 4\pi}{(2\pi\hbar)^3}(2mT)^{3/2}\frac{1}{2}\frac{1}{2}\pi^{1/2}\frac{1}{(3/2)(1/2)\pi^{1/2}}\int_0^\infty \frac{x^{3/2}}{1 + z^{-1}e^x}dx. \qquad (7.3.18)$$

Taking into account that $\Gamma(5/2) = (3/2)(1/2)\pi^{1/2}$ we can identify the function $f_{5/2}(z)$ in the previous expression and we obtain:

$$\frac{P\Omega}{T} = \sum_\varepsilon \ln(1 + ze^{-\beta\varepsilon}) = \Omega v \frac{(2\pi mT)^{3/2}}{(2\pi\hbar)^3} f_{5/2}(z). \tag{7.3.19}$$

If we eliminate the volume from both sides we have

$$\frac{P}{T} = v\frac{(2\pi mT)^{3/2}}{(2\pi\hbar)^3} f_{5/2}(z). \tag{7.3.20}$$

In the limit $z \to 0$ we have that $f_{5/2}(z) \to z$ and therefore,

$$\frac{P}{T} = v\frac{(2\pi mT)^{3/2}}{(2\pi\hbar)^3} e^{\mu/T}, \tag{7.3.21}$$

Taking into account that in this limit,

$$\mu = T\ln\left(\frac{\rho}{v}\left(\frac{2\pi\hbar^2}{mT}\right)^{3/2}\right) \implies e^{\mu/T} = \frac{\rho}{v}\left(\frac{2\pi\hbar^2}{mT}\right)^{3/2}, \tag{7.3.22}$$

one obtains

$$P = \rho T \tag{7.3.23}$$

which is the pressure as function of ρ and T for a classical ideal gas. The pressure is linear in temperature and the proportionality factor is just the density.

PROBLEM 7.4
Low T expansion

For a given density, at low temperatures $\eta = \mu/T \gg 0$. The Fermi integrals

$$J_\nu(\eta) = \int_0^\infty \frac{x^\nu}{1 + e^{x-\eta}}\, dx,$$

can be expanded for $\eta \gg 0$ as

$$\int_0^\infty \frac{\phi(x)}{1 + e^{x-\eta}}\, dx = \int_0^\eta \phi(x)\, dx + \frac{\pi^2}{6}\left(\frac{d\phi}{dx}\right)_{x=\eta} + \frac{7\pi^4}{360}\left(\frac{d^3\phi}{dx^3}\right)_{x=\eta} + \cdots$$

(a) Apply the previous expression to derive an expansion for the Fermi integrals $J_{1/2}(\eta)$ and $J_{3/2}(\eta)$ when $\eta \gg 0$.

(b) Using these results find a low-temperature expansion for the chemical potential, by inverting the relation

$$\rho = \frac{v}{2^{1/2}\pi^2}\left(\frac{mT}{\hbar^2}\right)^{3/2}\int_0^\infty \frac{x^{1/2}}{1 + e^{x-\eta}}\, dx, \tag{7.4.1}$$

where $\eta = \mu/T$.

(c) Using the expansion of μ derive a low-temperature expansion for the energy per particle,

$$e = \frac{1}{\rho} \frac{v}{2^{1/2}\pi^2} \left(\frac{mT}{\hbar^2}\right)^{3/2} T \int_0^\infty \frac{x^{3/2}}{1+e^{x-\eta}} dx. \tag{7.4.2}$$

(d) Derive the low-temperature expansion of the entropy per particle by using that the entropy can be written as

$$s(\rho,T) = \frac{5}{3} \frac{e(\rho,T)}{T} - \eta. \tag{7.4.3}$$

(e) Find also the low-temperature expansion of the free-energy per particle

$$f(\rho,T) = e(\rho,T) - Ts(\rho,T). \tag{7.4.4}$$

(f) Derive the low-temperature expansion for the pressure by using

$$P(\rho,T) = \rho^2 \frac{\partial f(\rho,T)}{\partial \rho}. \tag{7.4.5}$$

(g) Finally, check that the low-temperature expansion of $\mu(\rho,T)$ given by

$$\mu(\rho,T) = f(\rho,T) + \frac{P(\rho,T)}{\rho} \tag{7.4.6}$$

coincides with the low temperature expansion of $\mu(\rho,T)$ previously derived by inverting the relation between the density and the chemical potential.

(a) For a given density, at very low temperatures, $\eta = \mu/T \gg 0$; then, the Fermi integrals admit the following expansion:

$$\int_0^\infty \frac{\phi(x)}{1+e^{x-\eta}} dx = \int_0^\eta \phi(x)\, dx + \frac{\pi^2}{6}\left(\frac{d\phi}{dx}\right)_{x=\eta} + \frac{7\pi^4}{360}\left(\frac{d^3\phi}{dx}\right)_{x=\eta} + \cdots \tag{7.4.7}$$

In particular,

$$J_{1/2}(\eta) = \int_0^\infty \frac{x^{1/2}}{1+e^{x-\eta}} dx = \int_0^\eta x^{1/2} dx + \frac{\pi^2}{6}\frac{1}{2}x^{-1/2}\bigg|_{x=\eta} + \cdots$$

$$= \frac{2}{3}x^{3/2}\bigg|_0^\eta + \frac{\pi^2}{12}\frac{1}{x^{1/2}}\bigg|_{x=\eta} + \cdots = \frac{2}{3}\eta^{3/2} + \frac{\pi^2}{12}\frac{1}{\eta^{1/2}} + \cdots$$

$$= \frac{2}{3}\left(\frac{\mu}{T}\right)^{3/2}\left(1 + \frac{\pi^2}{8}\left(\frac{T}{\mu}\right)^2 + \cdots\right).$$

In a similar way,

$$J_{3/2}(\eta) = \int_0^\infty \frac{x^{3/2}}{1+e^{x-\eta}} = \int_0^\eta x^{3/2}dx + \frac{\pi^2}{6}\frac{3}{2}x^{1/2}\Big|_{x=\eta} + \cdots$$

$$= \frac{2}{5}x^{5/2}\Big|_0^\eta + \frac{\pi^2}{6}\frac{3}{2}\eta^{1/2} + \cdots = \frac{2}{5}\left(\frac{\mu}{T}\right)^{5/2}\left(1+\frac{5}{2}\frac{\pi^2}{6}\frac{3}{2}\left(\frac{T}{\mu}\right)^2+\cdots\right)$$

$$= \frac{2}{5}\left(\frac{\mu}{T}\right)^{5/2}\left(1+\frac{5\pi^2}{8}\left(\frac{T}{\mu}\right)^2+\cdots\right).$$

(b) At the quadratic order in $T/\mu \ll 1$,

$$\rho \simeq \frac{v}{2^{1/2}\pi^2}\left(\frac{m}{\hbar^2}\right)^{3/2}\frac{2}{3}\mu^{3/2}\left(1+\frac{\pi^2}{8}\left(\frac{T}{\mu}\right)^2+\cdots\right).$$

We should isolate μ, and find the quadratic correction in temperature to the zero-temperature chemical potential $\varepsilon_F(T=0)$ of the free Fermi sea,

$$\varepsilon_F(T=0) = \frac{\hbar^2}{2m}\left(\frac{6\pi^2\rho}{v}\right)^{2/3}. \tag{7.4.8}$$

Then,

$$\frac{\hbar^2}{2m}\left(\frac{6\pi^2\rho}{v}\right)^{2/3} \simeq \mu\left(1+\frac{\pi^2}{8}\left(\frac{T}{\mu}\right)^2\right)^{2/3}. \tag{7.4.9}$$

Using the expansion $(1+x)^{2/3} \simeq 1+2x/3$ valid when $x \to 0$, we get

$$\frac{\hbar^2}{2m}\left(\frac{6\pi^2\rho}{v}\right)^{2/3} \simeq \mu\left(1+\frac{\pi^2}{12}\left(\frac{T}{\mu}\right)^2\right). \tag{7.4.10}$$

We solve this equation, iteratively

$$\mu \simeq \varepsilon_F\frac{1}{1+\frac{\pi^2}{12}\left(\frac{T}{\mu}\right)^2} \simeq \varepsilon_F\left(1-\frac{\pi^2}{12}\left(\frac{T}{\mu}\right)^2\right), \tag{7.4.11}$$

valid when $T/\mu \ll 1$. Now we substitute again μ by ε_F, and get the expansion up to T^2

$$\mu \simeq \varepsilon_F - T^2\frac{\pi^2}{12\varepsilon_F} = \varepsilon_F\left(1-\frac{\pi^2}{12}\left(\frac{T}{\varepsilon_F}\right)^2\right). \tag{7.4.12}$$

(c) The energy per particle is given by

$$e = \frac{1}{\rho}\frac{v}{2^{1/2}\pi^2}\left(\frac{mT}{\hbar^2}\right)^{3/2}T\int_0^\infty \frac{x^{3/2}}{1+e^{x-\eta}}dx, \tag{7.4.13}$$

which for $\eta \gg 0$ is expanded as

$$e \simeq \frac{1}{\rho} \frac{v}{2^{1/2}\pi^2} \left(\frac{m}{\hbar^2}\right)^{3/2} T^{5/2} \frac{2}{5} \left(\frac{\mu}{T}\right)^{5/2} \left(1 + \frac{\pi^2 5}{8}\left(\frac{T}{\mu}\right)^2\right). \qquad (7.4.14)$$

Then using the expansion of μ up to second order in T/μ,

$$\mu \simeq \varepsilon_F \left(1 - \frac{\pi^2}{12}\left(\frac{T}{\varepsilon_F}\right)^2\right), \qquad (7.4.15)$$

we obtain

$$e \simeq \frac{1}{\rho} \frac{v}{2^{1/2}\pi^2} \left(\frac{m}{\hbar^2}\right)^{3/2} \frac{2}{5} \varepsilon_F^{5/2} \left(1 - \frac{\pi^2}{12}\left(\frac{T}{\varepsilon_F}\right)^2\right)^{5/2}$$

$$\times \left(1 + \frac{5\pi^2}{8}\left(\frac{T}{\varepsilon_F}\right)^2 \frac{1}{\left(1 - \frac{\pi^2}{12}\left(\frac{T}{\varepsilon_F}\right)^2\right)^2}\right).$$

Now, keeping quadratic terms in T/ε_F, we can compute the Taylor expansion of the T−dependent factors:

$$\left(1 - \frac{\pi^2}{12}\left(\frac{T}{\varepsilon_F}\right)^2\right)^{5/2} \simeq 1 - \frac{5\pi^2}{24}\left(\frac{T}{\varepsilon_F}\right)^2,$$

$$1 + \frac{5\pi^2}{8}\left(\frac{T}{\varepsilon_F}\right)^2 \simeq 1 + \frac{5\pi^2}{8}\left(\frac{T}{\varepsilon_F}\right)^2 \frac{1}{\left(1 - \frac{\pi^2}{12}\left(\frac{T}{\varepsilon_F}\right)^2\right)^2}$$

$$\simeq 1 + \frac{5\pi^2}{8}\left(\frac{T}{\varepsilon_F}\right)^2 \left(1 + \frac{\pi^2}{6}\left(\frac{T}{\varepsilon_F}\right)^2\right)$$

$$\simeq 1 + \frac{5\pi^2}{8}\left(\frac{T}{\varepsilon_F}\right)^2.$$

Now we need to get ρ to substitute it in the expression for e. We can obtain its expression using the relation between the Fermi momentum and the energy at $T = 0$ and substituting the degeneracy $v = 2$:

$$\varepsilon_F(T = 0) = \frac{\hbar^2}{2m}\left(\frac{6\pi^2\rho}{v}\right)^{2/3} \implies \rho = \frac{2}{3}\frac{1}{2^{1/2}\pi^2}\left(\frac{m}{\hbar}\right)^2 \varepsilon_F^{3/2}.$$

Bringing everything together,

$$e \simeq \frac{3}{5}\varepsilon_F \left(1 - \frac{5\pi^2}{24}\left(\frac{T}{\varepsilon_F}\right)^2\right)\left(1 + \frac{5\pi^2}{8}\left(\frac{T}{\varepsilon_F}\right)^2\right) \simeq \frac{3}{5}\varepsilon_F\left(1 + \frac{5\pi^2}{12}\left(\frac{T}{\varepsilon_F}\right)^2\right).$$

At low temperature, the energy per particle increases quadratically with T,

$$e(\rho,T) \simeq e(\rho,T=0) + aT^2, \tag{7.4.16}$$

with $a = \pi^2/(4\varepsilon_F)$.

(d) The entropy can be expressed as

$$s(\rho,T) = \frac{5}{3}\frac{e(\rho,T)}{T} - \eta. \tag{7.4.17}$$

Now using the expansion of each quantity we can write

$$s = \frac{5}{3}\frac{1}{T}\left[\frac{3}{5}\varepsilon_F\left(1 + \frac{5\pi^2}{12}\left(\frac{T}{\varepsilon_F}\right)^2\right)\right] - \frac{\varepsilon_F}{T}\left(1 - \frac{\pi^2}{12}\left(\frac{T}{\varepsilon_F}\right)^2\right)$$

$$= \frac{\varepsilon_F}{T} + \frac{5\pi^2}{12}\frac{T}{\varepsilon_F} - \frac{\varepsilon_F}{T} + \frac{\pi^2}{12}\frac{T}{\varepsilon_F} = \frac{1}{2}\pi^2\frac{T}{\varepsilon_F},$$

which is linear in T.

(e) The free energy is expressed as

$$f(\rho,T) = e(\rho,T) - Ts(\rho,T). \tag{7.4.18}$$

Then using the expansion of each of term we have:

$$f(\rho,T) \simeq \frac{3}{5}\varepsilon_F\left(1 + \frac{5\pi^2}{12}\left(\frac{T}{\varepsilon_F}\right)^2\right) - T\frac{\pi^2}{2}\frac{T}{\varepsilon_F}$$

$$= \frac{3}{5}\varepsilon_F - \frac{\pi^2}{4}\frac{T^2}{\varepsilon_F} = \frac{3}{5}\varepsilon_F\left(1 - \frac{5\pi^2}{12}\left(\frac{T}{\varepsilon_F}\right)^2\right),$$

which decreases quadratically with T. Actually it can be expressed as $f(\rho,T) \simeq e(\rho,T=0) - aT^2$ where a is the same factor that appears in the low temperature expansion of $e(\rho,T)$, i.e. $a = \pi^2/(4\varepsilon_F)$.

(f) Doing a similar process for the pressure

$$P(\rho,T) = \rho^2\frac{\partial f(\rho,T)}{\partial\rho}, \tag{7.4.19}$$

one gets

$$P(\rho,T) = \rho^2\frac{3}{5}\frac{\partial\varepsilon_F}{\partial\rho} - \frac{\pi^2}{4}T^2\rho^2\frac{\partial}{\partial\rho}\left(\frac{1}{\varepsilon_F}\right) = \frac{2}{5}\rho\varepsilon_F + \rho\frac{\pi^2}{6}\frac{T^2}{\varepsilon_F}$$

$$= \frac{2}{5}\rho\varepsilon_F\left(1 + \frac{5\pi^2}{12}\left(\frac{T}{\varepsilon_F}\right)^2\right).$$

The pressure increases quadratically with T.

(g) Finally, the chemical potential can be also derived as

$$\mu(\rho,T) = f(\rho,T) + \frac{P(\rho,T)}{\rho} .$$

Then, introducing the expansion of each quantity we have

$$\mu(\rho,T) \simeq \frac{3}{5}\varepsilon_F \left(1 - \frac{5\pi^2}{12}\left(\frac{T}{\varepsilon_F}\right)^2\right) + \frac{2}{5}\varepsilon_F + \frac{\pi^2}{6}\frac{T^2}{\varepsilon_F}$$

$$= \varepsilon_F \left(1 - \frac{\pi^2}{12}\left(\frac{T}{\varepsilon_F}\right)^2\right),$$

which coincides with the expression of μ obtained from the normalization of the density.

PROBLEM 7.5
Partition function of the free Fermi system

Consider a free Fermi system described by the Hamiltonian

$$\hat{H}_0 = \sum_k \varepsilon_k \hat{c}_k^\dagger \hat{c}_k . \tag{7.5.1}$$

The Fock states associated to this Hamiltonian are described as

$$|\phi_i\rangle = |n_1^i, n_2^i, \ldots n_k^i, \ldots\rangle , \tag{7.5.2}$$

with $n_k^i = 1$ or 0, depending on whether the single-particle state k is occupied or not, respectively.

These states are eigenstates of $\hat{H}_0$, $\hat{H}_0|\phi_i^0\rangle = E_0^i|\phi_i^0\rangle$ with $E_0^i = \sum_k n_k^i \varepsilon_k$. Notice that we are working in the grand canonical ensemble and $\sum_k n_k^i = N^i$ is the number of particles in each state. The number of particles is not fixed. The distribution operator is defined as

$$\hat{\rho}_0 = e^{-\beta(\hat{H}_0 - \mu\hat{N})} , \tag{7.5.3}$$

where $\beta = 1/T$, i.e., the inverse of the temperature, and μ is the chemical potential.

(a) Calculate the grand partition function $Z_0 = \mathrm{Tr}(\hat{\rho}_0)$.

(b) Calculate the thermal average of the occupation number operator, $\hat{c}_k^\dagger \hat{c}_k$, as a function of μ and T in the grand canonical ensemble.

(a) The matrix elements of $\hat{\rho}_0$ in the Fock states are:

$$\rho_{0,i} = \langle \phi_i | \hat{\rho}_0 | \phi_i \rangle = \left\langle n_1^i, n_2^i, \ldots, n_k^i, \ldots \left| e^{-\beta \Sigma_k (\varepsilon_k - \mu) \hat{c}_k^\dagger \hat{c}_k} \right| n_1^i, n_2^i, \ldots, n_k^i, \ldots \right\rangle.$$

(7.5.4)

Notice that $\hat{c}_k^\dagger \hat{c}_k | n_1^i, \ldots, n_k^i, \ldots \rangle = n_k^i | n_1^i, \ldots, n_k^i, \ldots \rangle$.

Therefore,

$$\rho_{0,i} = \langle \phi_i | \hat{\rho}_0 | \phi_i \rangle = e^{-\beta \Sigma_k (\varepsilon_k - \mu) n_k^i} = \prod_k e^{-\beta (\varepsilon_k - \mu) n_k^i}.$$

(7.5.5)

With the matrix elements, one can calculate the trace:

$$Z_0 = \text{Tr}(\hat{\rho}_0) = \sum_i \rho_{0,i}.$$

(7.5.6)

This summation is equivalent to:

$$\sum_i \rho_{0,i} = \sum_i \prod_k e^{-\beta (\varepsilon_k - \mu) n_k^i}.$$

In this summation, the first term $i = 1$ corresponds to the vacuum: $\{n_1^i, \ldots, n_k^i, \ldots\} = \{0, 0, 0, \ldots\}$; the next terms $i = 2, 3, \ldots$ correspond to the states with one particle: $\{1, 0, 0, \ldots\}, \{0, 1, 0, \ldots\}, \{0, 0, 1, \ldots\}, \ldots$; then, the states with two particles, and so on.

Finally, the summation is given by:

$$Z_0 = \sum_i \rho_{0,i} = 1 + e^{-\beta (\varepsilon_1 - \mu)} + e^{-\beta (\varepsilon_2 - \mu)} + \cdots$$

$$+ e^{-\beta (\varepsilon_1 - \mu)} e^{-\beta (\varepsilon_2 - \mu)} + e^{-\beta (\varepsilon_1 - \mu)} e^{-\beta (\varepsilon_3 - \mu)} + \cdots$$

$$= \prod_k \left(1 + e^{-\beta (\varepsilon_k - \mu)} \right).$$

(b) Having Z_0, one can calculate the thermal average of any operator, in particular of $\hat{c}_k^\dagger \hat{c}_k$, as

$$\left\langle \hat{c}_k^\dagger \hat{c}_k \right\rangle_0 = \frac{\text{Tr}(\hat{c}_k^\dagger \hat{c}_k \hat{\rho}_0)}{\text{Tr}(\hat{\rho}_0)}.$$

(7.5.7)

In the free system, $\hat{\rho}_0$ is diagonal in the Fock basis, and $\hat{c}_k^\dagger \hat{c}_k$ is also diagonal. Therefore

$$\left\langle \hat{c}_k^\dagger \hat{c}_k \right\rangle_0 = \sum_i \left\langle \phi_i \left| \hat{c}_k^\dagger \hat{c}_k \right| \phi_i \right\rangle \frac{\langle \phi_i | \hat{\rho}_0 | \phi_i \rangle}{Z_0}$$

(7.5.8)

and

$$\frac{\rho_{0,i}}{Z_0} = \frac{\langle \phi_i | \hat{\rho}_0 | \phi_i \rangle}{Z_0} = \frac{\prod_l e^{-\beta (\varepsilon_l - \mu) n_l^i}}{\prod_l (1 + e^{-\beta (\varepsilon_l - \mu)})} = \prod_l \frac{e^{-\beta (\varepsilon_l - \mu) n_l^i}}{(1 + e^{-\beta (\varepsilon_l - \mu)})}.$$

(7.5.9)

Then,

$$\left\langle \hat{c}_k^\dagger \hat{c}_k \right\rangle_0 = \sum_i \left\langle n_1^i, \ldots n_k^i, \ldots \left| c_k^\dagger c_k \right| n_1^i, \ldots, n_k^i, \ldots \right\rangle \prod_l \frac{e^{-\beta(\varepsilon_l - \mu) n_l^i}}{(1 + e^{-\beta(\varepsilon_l - \mu)})}. \quad (7.5.10)$$

In the summation, we have contribution only from the states with $n_k = 1$, and all of them contain the factor

$$\frac{e^{-\beta(\varepsilon_k - \mu)}}{1 + e^{-\beta(\varepsilon_k - \mu)}}. \quad (7.5.11)$$

Therefore, we can write

$$\left\langle \hat{c}_k^\dagger \hat{c}_k \right\rangle_0 = \frac{e^{-\beta(\varepsilon_k - \mu)}}{1 + e^{-\beta(\varepsilon_k - \mu)}} \sum_i \prod_{l \neq k} \frac{e^{-\beta(\varepsilon_l - \mu) n_l^i}}{1 + e^{-\beta(\varepsilon_l - \mu)}}. \quad (7.5.12)$$

Now,

$$\frac{\sum_i \prod_{l \neq k} e^{-\beta(\varepsilon_l - \mu) n_l^i}}{\prod_{l \neq k} \left[1 + e^{-\beta(\varepsilon_k - \mu)} \right]} = \frac{\prod_{l \neq k} \left[1 + e^{-\beta(\varepsilon_k - \mu)} \right]}{\prod_{l \neq k} \left[1 + e^{-\beta(\varepsilon_k - \mu)} \right]} = 1. \quad (7.5.13)$$

Finally,

$$\left\langle \hat{c}_k^\dagger \hat{c}_k \right\rangle_0 = \frac{e^{-\beta(\varepsilon_k - \mu)}}{1 + e^{-\beta(\varepsilon_k - \mu)}} = \frac{1}{1 + e^{\beta(\varepsilon_k - \mu)}}. \quad (7.5.14)$$

In the case of the free Fermi sea, $\varepsilon_k = \hbar^2 k^2 / (2m)$, where $\vec{k}$ is the momentum of the single-particle state. Usually, for a given ρ and $\beta = 1/T$, one determines μ by inverting the relation

$$\rho = \frac{v}{(2\pi)^3} \int d\vec{k} \frac{1}{1 + e^{\beta(\varepsilon_k - \mu)}}. \quad (7.5.15)$$

Alternatively, for a given μ, we can calculate ρ.

PROBLEM 7.6
Entropy of the non-interacting Fermi sea

Consider a free Fermi system described by the Hamiltonian

$$\hat{H} = \sum_k \varepsilon_k \hat{c}_k^\dagger \hat{c}_k. \quad (7.6.1)$$

The grand canonical potential for the system is given by

$$\Omega = -T \ln Z, \quad (7.6.2)$$

where Z is the partition function defined by:

$$Z(\mu, V, T) = \text{Tr}\left(e^{-\beta(\hat{H} - \mu \hat{N})}\right), \tag{7.6.3}$$

where μ is the chemical potential, $\hat{N} = \sum_{\vec{k}} \hat{a}_{\vec{k}}^{\dagger} \hat{a}_{\vec{k}}$ is the number operator, $\beta = 1/T$ and V is the volume.

Show that the entropy, which is defined like

$$S = -\left(\frac{\partial \Omega}{\partial T}\right)_{\mu, V}, \tag{7.6.4}$$

can be written as

$$S = -\sum_{\vec{k}} \{ f(\varepsilon) \ln f(\varepsilon) + (1 - f(\varepsilon)) \ln(1 - f(\varepsilon)) \}, \tag{7.6.5}$$

where

$$f(\varepsilon) = \frac{1}{1 + e^{(\varepsilon - \mu)/T}}. \tag{7.6.6}$$

The partition function for the free Fermi sea can be expressed as

$$Z = \prod_{\vec{k}} \left(1 + e^{-(\varepsilon(k) - \mu)/T}\right). \tag{7.6.7}$$

Therefore the grand canonical potential is written as

$$\Omega = -T \sum_{\vec{k}} \ln \left[1 + e^{-(\varepsilon(k) - \mu)/T}\right], \tag{7.6.8}$$

and the entropy $S = -(\partial \Omega / \partial T)_{\mu, V}$ has several contributions,

$$S = \sum_{\vec{k}} \left\{ \ln \left[1 + e^{-(\varepsilon(k) - \mu)/T}\right] + T \frac{e^{-(\varepsilon(k) - \mu)/T}}{1 + e^{-(\varepsilon(k) - \mu)/T}} \frac{(\varepsilon(k) - \mu)}{T^2} \right\}, \tag{7.6.9}$$

which can be transformed to

$$S = \sum_{\vec{k}} \left\{ \ln \left[\frac{e^{(\varepsilon(k) - \mu)/T}}{e^{(\varepsilon(k) - \mu)/T}} \left[1 + e^{-(\varepsilon(k) - \mu)/T}\right] \right] + \frac{\varepsilon(k) - \mu}{T} \frac{1}{1 + e^{(\varepsilon(k) - \mu)/T}} \right\}, \tag{7.6.10}$$

then

$$S = \sum_{\vec{k}} \left\{ \ln \left[e^{(\varepsilon(k) - \mu)/T} + 1 \right] - \ln \left[e^{(\varepsilon(k) - \mu)/T} \right] + \frac{\varepsilon(k) - \mu}{T} \frac{1}{1 + e^{(\varepsilon(k) - \mu)/T}} \right\}. \tag{7.6.11}$$

From now on we will not specify the argument in $\varepsilon(k)$, but just write ε. Then we have

$$S = -\sum_{\vec{k}} \left\{ -\ln\left[1 + e^{(\varepsilon-\mu)/T}\right] + \frac{\varepsilon-\mu}{T}\left[1 - \frac{1}{1+e^{(\varepsilon-\mu)/T}}\right] \right\}. \qquad (7.6.12)$$

But,

$$\frac{\varepsilon-\mu}{T} = \ln\left(1 - \frac{1}{1+e^{(\varepsilon-\mu)/T}}\right) + \ln\left(1 + e^{(\varepsilon-\mu)/T}\right). \qquad (7.6.13)$$

Then,

$$S = -\sum_{\vec{k}} \left\{ -\ln\left[1 + e^{(\varepsilon-\mu)/T}\right] + \left[\ln\left(1 - \frac{1}{1+e^{(\varepsilon-\mu)/T}}\right) + \ln\left(1 + e^{(\varepsilon-\mu)/T}\right)\right] \right.$$
$$\left. \times \left[1 - \frac{1}{1+e^{(\varepsilon-\mu)/T}}\right] \right\},$$

which can be grouped as

$$S = -\sum_{\vec{k}} \left\{ \ln\left(e^{(\varepsilon-\mu)/T} + 1\right)\left(-1 + 1 - \frac{1}{1+e^{(\varepsilon-\mu)/T}}\right) \right.$$
$$\left. + \left(1 - \frac{1}{1+e^{(\varepsilon-\mu)/T}}\right)\ln\left(1 - \frac{1}{1+e^{(\varepsilon-\mu)/T}}\right) \right\}$$

and written as

$$S = -\sum_{\vec{k}} \left\{ \frac{1}{1+e^{(\varepsilon-\mu)/T}} \ln\left(\frac{1}{1+e^{(\varepsilon-\mu)/T}}\right) \right.$$
$$\left. + \left(1 - \frac{1}{1+e^{(\varepsilon-\mu)/T}}\right)\ln\left(1 - \frac{1}{1+e^{(\varepsilon-\mu)/T}}\right) \right\}.$$

Finally, if we introduce the Fermi thermal occupation function

$$f(\varepsilon) = \frac{1}{1+e^{(\varepsilon-\mu)/T}}, \qquad (7.6.14)$$

and taking into account the spin degeneracy ν, we have

$$S = -\nu \sum_{\vec{k}} \left\{ f(\varepsilon)\ln f(\varepsilon) + (1 - f(\varepsilon))\ln(1 - f(\varepsilon)) \right\}. \qquad (7.6.15)$$

If we transform the discrete sum to the continuum we have:

$$\sum_{\vec{k}} = \frac{V}{(2\pi)^3} \int d\vec{k}. \qquad (7.6.16)$$

Finally, we get:

$$S = -\frac{\nu V}{(2\pi)^3} \int d\vec{k} \left\{ f(\varepsilon)\ln f(\varepsilon) + (1 - f(\varepsilon))\ln(1 - f(\varepsilon)) \right\}. \qquad (7.6.17)$$

8 Distribution functions

Correlation is a crucial concept in science. It is deeply related to our understanding of the interrelation between different parties or physical systems, and thus to the casual connection between phenomena. In physics, correlations appear, usually but not only, as a consequence of the interaction between different parts of a system.

In usual statistics two variables are said to be correlated if their outcomes depend linearly on each other. In our case, more physics motivated, two systems are correlated if their outcomes are not independent of each other. Interestingly, in many-body quantum physics, there are non-trivial correlations that appear already in non-interacting fermionic systems. This is due to the Pauli exclusion principle, that builds a strong correlation on the measurement of, for instance, the positions of two identical fermions.

To quantify the number and properties of the correlations we can use so-called distribution functions . In this chapter we present three exercises about the distribution functions of the Fermi sea.

PROBLEM 8.1
Sequential condition for the two-body distribution function of the free Fermi sea

Consider the two-body distribution function , $g(r)$, of a free Fermi sea, given by the expression

$$g(r) = 1 - \frac{l(k_F r)^2}{\nu}, \tag{8.1.1}$$

where $\nu = 2(4)$ is the spin(-isospin) degeneracy and the Slater function $l(k_F r)$ is expressed in terms of Bessel functions

$$l(k_F r) = \frac{\nu}{(2\pi)^3 \rho} \int_{|\vec{k}| < k_F} d\vec{k}\, e^{i\vec{k}\vec{r}} = \frac{3 j_1(k_F r)}{k_F r}.$$

Here, the Bessel function $j_1(k_F r)$ is given by the expression

$$j_1(k_F r) = \frac{\sin(k_F r)}{(k_F r)^2} - \frac{\cos(k_F r)}{k_F r}.$$

Show that the sequential condition

$$\rho \int d\vec{r}\, (g(r) - 1) = -1 \tag{8.1.2}$$

is fulfilled.

(a) The two-body distribution function of the Fermi sea reads

$$g(r) = 1 - \frac{l(k_F r)^2}{\nu}, \tag{8.1.3}$$

For the purpose of this exercise, it is convenient to learn how to perform two radial integrals over $l(k_F r)$ and $l^2(k_F r)$. Starting with the one for $l(k_F r)$, we find:

$$\int d\vec{r}\, l(k_F r) = \frac{\nu}{(2\pi)^3 \rho} \int d\vec{r} \int_{|\vec{k}| < k_F} d\vec{k}\, e^{i\vec{k}\vec{r}}$$

$$= \frac{\nu}{\rho} \int_{|\vec{k}| < k_F} d\vec{k}\, \frac{1}{(2\pi)^3} \int d\vec{r}\, e^{i\vec{k}\vec{r}} = \frac{\nu}{\rho} \int_{|\vec{k}| < k_F} d\vec{k}\, \delta(\vec{k}) = \frac{\nu}{\rho}.$$

Similarly, the integral for $l^2(k_F r)$ reads:

$$\int d\vec{r}\, l^2(k_F r) = \int d\vec{r}\, \frac{\nu}{(2\pi)^3} \frac{\nu}{(2\pi)^3} \left[\int_{|\vec{k}| < k_F} d\vec{k}\, e^{i\vec{k}\vec{r}} \right] \left[\int_{|\vec{k}'| < k_F} d\vec{k}'\, e^{i\vec{k}'\vec{r}} \right].$$

To perform this integral, we change the order of integration,

$$\int_{|\vec{k}|<k_F} d\vec{k} \int_{|\vec{k}'|<k_F} d\vec{k}' \frac{v^2}{\rho^2(2\pi)^3} \frac{1}{(2\pi)^3} \int d\vec{r}\, e^{i(\vec{k}+\vec{k}')\vec{r}}$$

$$= \int_{|\vec{k}|<k_F} d\vec{k} \int_{|\vec{k}'|<k_F} d\vec{k}' \frac{v^2}{\rho^2(2\pi)^3} \delta(\vec{k}+\vec{k}') = \frac{v^2}{\rho^2(2\pi)^3} \int_{|\vec{k}|<k_F} d\vec{k} = \frac{v}{\rho}.$$

With this expression and, using Eq. (8.1.1), we find

$$\rho \int d\vec{r}\, (g(r)-1) = \rho \int d\vec{r}\left(-\frac{l(k_F r)^2}{v}\right) = -\frac{\rho}{v}\frac{v}{\rho} = -1.$$

PROBLEM 8.2
Spin components of the two-body distribution function

The expectation value of a two-body operator

$$\hat{V} = \sum_{i<j} v(r_{ij})\hat{O}_{ij} \tag{8.2.1}$$

is given by

$$\frac{<\hat{V}>}{N} = \frac{1}{2}\rho \int d\vec{r}\, v(r)\left\{\mathrm{Tr}\left[\hat{O}_{12}\mathscr{G}(\vec{r}_1\vec{\sigma}_1,\vec{r}_2\vec{\sigma}_2)\right]\right\} \tag{8.2.2}$$

where T_r indicates the trace in the spin-space and

$$\mathscr{G}(\vec{r}_1\vec{\sigma}_1;\vec{r}_2\vec{\sigma}_2) = \frac{N(N-1)}{\rho^2} \int \Psi^*(x_1,\ldots,x_N)\Psi(x_1,\ldots,x_N)dx_3\ldots dx_N \tag{8.2.3}$$

where integration over x_i implies both the integration over the spatial coordinates and a summation over the spin variables $\vec{\sigma}_i$. Then the distribution function $\mathscr{G}(\vec{r}_1\vec{\sigma}_1;\vec{r}_2\vec{\sigma}_2)$ can be expressed in several ways:

$$\begin{aligned}
\mathscr{G}(\vec{r}_1\vec{\sigma}_1;\vec{r}_2\vec{\sigma}_2) &= \frac{1}{v^2}\left[g_c(r_{12}) + (\vec{\sigma}_1\vec{\sigma}_2)g_{\sigma_1\sigma_2}(r_{12})\right] \\
&= \frac{1}{v^2}\left[\hat{P}_s g_s(r_{12}) + \hat{P}_T g_T(r_{12})\right] \\
&= \frac{1}{v^2}\left[g_0(r_{12}) + \hat{P}_{12} g_M(r_{12})\right],
\end{aligned} \tag{8.2.4}$$

where $\hat{P}_s$ and $\hat{P}_t$ are the singlet and triplet spin projection operators, and $\hat{P}_{12}$ is the spin exchange operator.

(a) Find out the relation between the different spin components of the distribution function.

(b) Calculate explicitly the different components for the free Fermi sea, with a Fermi momentum k_F and spin degeneracy two.

(a) To obtain the different component form $\mathscr{G}(\vec{r}_1\vec{\sigma}_1;\vec{r}_2\vec{\sigma}_2)$ we should calculate some traces in the spin space. For instance,

$$g_c(r_{12}) = \text{Tr}\left[\mathscr{G}(\vec{r}_1\vec{\sigma}_1;\vec{r}_2\vec{\sigma}_2)\right], \qquad (8.2.5)$$

and

$$g_{\sigma_1\sigma_2} = \frac{1}{3}\text{Tr}\left[(\vec{\sigma}_1\vec{\sigma}_2)\mathscr{G}(\vec{r}_1\vec{\sigma}_1;\vec{r}_2\vec{\sigma}_2)\right]. \qquad (8.2.6)$$

In fact, if we take

$$\mathscr{G}(\vec{r}_1\vec{\sigma}_1;\vec{r}_2\vec{\sigma}_2) = \frac{1}{v^2}\left[g_c(r_{12}) + (\vec{\sigma}_1\vec{\sigma}_2)g_{\sigma_1\sigma_2}(r_{12})\right], \qquad (8.2.7)$$

and taking into account that $\text{Tr}(I) = v^2$ and $\text{Tr}(\vec{\sigma}_1\vec{\sigma}_2) = 0$ we easily obtain the first relation. Next, we consider

$$\text{Tr}\left[(\vec{\sigma}_1\vec{\sigma}_2)\mathscr{G}(\vec{r}_1\vec{\sigma}_1;\vec{r}_2\vec{\sigma}_2)\right] = \frac{1}{v^2}\text{Tr}\left[(\vec{\sigma}_1\vec{\sigma}_2)g_c(r_{12}) + (3 - 2(\vec{\sigma}_1\vec{\sigma}_2))g_{\sigma_1\sigma_2}(r_{12})\right]. \qquad (8.2.8)$$

Therefore,

$$\text{Tr}[(\vec{\sigma}_1\vec{\sigma}_2)\mathscr{G}(\vec{r}_1,\vec{\sigma}_1;\vec{r}_2\vec{\sigma}_2)] = \frac{3v^2}{v^2}g_{\sigma_1\sigma_2}(r_{12}). \qquad (8.2.9)$$

Then if we want to find the relation between two representations we should just use these relations but with the other representation. For instance, we know that

$$g_c(r_{12}) = \text{Tr}[\mathscr{G}(\vec{r}_1\vec{\sigma}_1,\vec{r}_2\vec{\sigma}_2)], \qquad (8.2.10)$$

but now we express $\mathscr{G}(\vec{r}_1\vec{\sigma}_1,\vec{r}_2\vec{\sigma}_2)$ in terms of $g_0(r_{12})$ and $g_M(r_{12})$,

$$g_c(r_{12}) = \frac{1}{v^2}\text{Tr}\left[g_0(r_{12}) + \frac{1}{2}(1 + \vec{\sigma}_1\vec{\sigma}_2)g_M(r_{12})\right] = g_0(r_{12}) + \frac{1}{2}g_M(r_{12}). \qquad (8.2.11)$$

On the other hand,

$$g_{\vec{\sigma}_1\vec{\sigma}_2}(r_{12}) = \frac{1}{3}\text{Tr}\left[(\vec{\sigma}_1\vec{\sigma}_2)\mathscr{G}(\vec{r}_1\vec{\sigma}_1,\vec{r}_2\vec{\sigma}_2)\right]. \qquad (8.2.12)$$

Now, we use that $\hat{P}_{12} = \frac{1}{2}(1 + \vec{\sigma}_1\vec{\sigma}_2)$ to transform

$$\vec{\sigma}_1\vec{\sigma}_2\hat{P}_{12} = \frac{\vec{\sigma}_1\vec{\sigma}_2}{2} + \frac{3 - 2(\vec{\sigma}_1\vec{\sigma}_2)}{2}, \qquad (8.2.13)$$

and

$$\text{Tr}[(\vec{\sigma}_1\vec{\sigma}_2)\hat{P}_{12}] = \frac{3}{2}v^2. \qquad (8.2.14)$$

Therefore,

$$g_{\vec{\sigma}_1\vec{\sigma}_2}(r_{12}) = \frac{1}{3}\frac{1}{v^2}\text{Tr}\left[(\vec{\sigma}_1\vec{\sigma}_2)\left(g_0(r_{12}) + \hat{P}_{12}g_M(r_{12})\right)\right] \qquad (8.2.15)$$

$$= \frac{1}{3}\frac{1}{v^2}v^2\frac{3}{2}g_M(r_{12}) = \frac{1}{2}g_M(r_{12}). \qquad (8.2.16)$$

Therefore

$$g_c(r_{12}) = g_0(r_{12}) + \frac{1}{2}g_M(r_{12}),$$

$$g_{\vec{\sigma}_1\vec{\sigma}_2}(r_{12}) = \frac{1}{2}g_M(r_{12}),$$ (8.2.17)

and

$$g_0(r_{12}) = g_c(r_{12}) - g_{\vec{\sigma}_1\vec{\sigma}_2}(r_{12}),$$

$$g_M(r_{12}) = 2\,g_{\vec{\sigma}_1\vec{\sigma}_2}(r_{12}).$$ (8.2.18)

Now we establish the relations with $g_S(r_{12})$ and $g_T(r_{12})$,

$$g_c(r_{12}) = \text{Tr}\left[\mathscr{G}(\vec{r}_1\vec{\sigma}_1, \vec{r}_2\vec{\sigma}_2))\right] = \frac{1}{v^2}\left\{\text{Tr}[\hat{P}_S]g_S(r_{12}) + \text{Tr}[\hat{P}_T]g_T(r_{12})\right\}. \quad (8.2.19)$$

Now we use that $\text{Tr}(\hat{P}_s) = v^2/4$ and that $\text{Tr}(\hat{P}_T) = 3/4v^2$, to get that

$$g_c(r_{12}) = \frac{1}{4}g_S(r_{12}) + \frac{3}{4}g_T(r_{12}),$$ (8.2.20)

while

$$g_{\vec{\sigma}_1\vec{\sigma}_2}(r_{12}) = \frac{1}{3}\text{Tr}[(\vec{\sigma}_1\vec{\sigma}_2)\mathscr{G}(\vec{r}_1\vec{\sigma}_1, \vec{r}_2\vec{\sigma}_2))]$$

$$= \frac{1}{3}\frac{1}{v^2}\text{Tr}\left[(\vec{\sigma}_1\vec{\sigma}_2)\hat{P}_S g_S(r_{12}) + (\vec{\sigma}_1\vec{\sigma}_2)\hat{P}_T g_T(r_{12})\right]. \quad (8.2.21)$$

Now we use $\hat{P}_S = \frac{1}{4}(1 - \vec{\sigma}_1\vec{\sigma}_2)$ and $\hat{P}_T = \frac{3}{4}(1 + \vec{\sigma}_1\vec{\sigma}_2)$ to simplify,

$$(\vec{\sigma}_1\vec{\sigma}_2)\hat{P}_S = \frac{3}{4}(\vec{\sigma}_1\vec{\sigma}_2) - 1),\quad (\vec{\sigma}_1\vec{\sigma}_2)\hat{P}_T = \frac{3}{4}(1 - (\vec{\sigma}_1\vec{\sigma}_2)).$$ (8.2.22)

Performing the traces we get

$$g_{\vec{\sigma}_1\vec{\sigma}_2}(r_{12}) = -\frac{1}{4}g_S(r_{12}) + \frac{1}{4}g_T(r_{12})$$ (8.2.23)

and we can also express the inverted relation:

$$g_T(r_{12}) = g_c(r_{12}) + g_{\vec{\sigma}_1\vec{\sigma}_2}(r_{12}),\quad g_S(r_{12}) = g_c(r_{12}) - 3\,g_{\vec{\sigma}_1\vec{\sigma}_2}(r_{12})$$ (8.2.24)

Finally we can also give the relations with $g_0(r_{12})$ and $g_M(r_{12})$, which would be obtained in a similar way:

$$g_T(r_{12}) = g_0(r_{12}) + g_M(r_{12}),\quad g_S(r_{12}) = g_0(r_{12}) - g_M(r_{12}).$$ (8.2.25)

(b) In particular for the unpolarized free Fermi sea, we can start from

$$\mathscr{G}_{FS}(\vec{r}_1\vec{\sigma}_1, \vec{r}_2\vec{\sigma}_2) = \frac{1}{v^2}\left[g_0(r_{12}) + P_{12}g_M(r_{12})\right],$$ (8.2.26)

with $v = 2$, $g_0(r_{12}) = 1$ and $g_M(r_{12}) = -l^2(k_F r)$, where $l^2(k_F r)$ is the square of the Slater function and k_F is the Fermi momentum. Then we have

$$g_c(r_{12}) = 1 - \frac{l^2(k_F r)}{v} \quad , \quad g_{\vec{\sigma}_1 \vec{\sigma}_2}(r_{12}) = -\frac{1}{v} l^2(k_F r), \qquad (8.2.27)$$

and

$$g_T(r_{12}) = 1 - l^2(k_F r) \quad , \quad g_S(r_{12}) = 1 + l^2(k_F r). \qquad (8.2.28)$$

PROBLEM 8.3
Spin of the free Fermi sea

Taking into account that the total spin of a system is

$$\vec{S} = \sum_i \vec{s}_i = 1/2 \sum_i \vec{\sigma}_i \qquad (8.3.1)$$

calculate $\vec{S}^2$ for the free Fermi sea.

(a) The total spin operator in units of $\hbar$ for an N particle system is

$$\vec{S} = \sum_i \vec{s}_i = \frac{1}{2} \sum_i \vec{\sigma}_i , \qquad (8.3.2)$$

then

$$\vec{S}^2 = \frac{1}{4} \sum_i \vec{\sigma}_i^2 + \frac{1}{4} 2 \sum_{i<j} \vec{\sigma}_i \vec{\sigma}_j , \qquad (8.3.3)$$

Taking into account that $\vec{\sigma}_i^2 = 3$ (three times the identity operator), we can write

$$\langle \phi_{FS} | \vec{S}^2 | \phi_{FS} \rangle = \langle \phi_{FS} | \frac{1}{4} \sum_i \vec{\sigma}_i^2 + \frac{1}{4} 2 \sum_{i<j} \vec{\sigma}_i \vec{\sigma}_j | \phi_{FS} \rangle$$

$$= \langle \phi_{FS} | \frac{3}{4} N | \phi_{FS} \rangle + \langle \phi_{FS} | \frac{1}{4} 2 \sum_{i<j} \vec{\sigma}_i \vec{\sigma}_j | \phi_{FS} \rangle . \qquad (8.3.4)$$

To evaluate the two-body part, we will use the two-body distribution function of the free Fermi sea:

$$\frac{1}{N} \langle \phi_{FS} | \frac{1}{4} 2 \sum_{i<j} \vec{\sigma}_i \vec{\sigma}_j | \phi_{FS} \rangle = \frac{1}{4} 2 \frac{1}{2} \rho \int d\vec{r} \, \text{Tr} \left[(\vec{\sigma}_1 \vec{\sigma}_2) \mathscr{G}(\vec{r}_1 \vec{\sigma}_1, \vec{r}_2 \vec{\sigma}_2) \right] , \qquad (8.3.5)$$

where ρ is the density of the free Fermi sea and Tr indicates the trace operation in the spin space of a two-particle system. The distribution function in given by:

$$\mathscr{G}(\vec{r}_1 \vec{\sigma}_1, \vec{r}_2 \vec{\sigma}_2) = \frac{1}{v^2} \left[g_c(r_{12}) + (\vec{\sigma}_1 \vec{\sigma}_2) \, g_{\vec{\sigma}_1 \vec{\sigma}_2}(r_{12}) \right] , \qquad (8.3.6)$$

where $v = 2$ is the spin degeneracy. Now we can use that

$$(\vec{\sigma}_1 \vec{\sigma}_2)^2 = 3 - 2(\vec{\sigma}_1 \vec{\sigma}_2), \tag{8.3.7}$$

and that

$$\text{Tr}(\vec{\sigma}_1 \vec{\sigma}_2) = 0, \quad \text{Tr}[(\vec{\sigma}_1 \vec{\sigma}_2)^2] = \text{Tr}(3) - 2\text{Tr}(\vec{\sigma}_1 \vec{\sigma}_2) = 3v^2 = 12, \tag{8.3.8}$$

and that for the free Fermi sea

$$\mathcal{G}(\vec{r}_1 \vec{\sigma}_1, \vec{r}_2 \vec{\sigma}_2) = \frac{1}{v^2}\left[(1 - \frac{l(k_F r)^2}{v})\hat{I} - \frac{l(k_F r)^2}{v}(\vec{\sigma}_1 \vec{\sigma}_2)\right], \tag{8.3.9}$$

where $l(k_F r)$ is the Slater function. Then,

$$\frac{1}{N}\langle \phi_{FS}| \frac{1}{4}2\sum_{i<j}\vec{\sigma}_i \vec{\sigma}_j |\phi_{FS}\rangle = \frac{1}{4}2\frac{1}{2}\frac{1}{v^2}\rho \int d^3r \, \text{Tr}((\vec{\sigma}_1\vec{\sigma}_2)(\vec{\sigma}_1\vec{\sigma}_2))g_{\vec{\sigma}_1\vec{\sigma}_2}(r)$$

$$= \frac{1}{4}2\frac{1}{2}\rho\frac{1}{v^2}v^2 3 \int d\vec{r} \, g_{\vec{\sigma}_1\vec{\sigma}_2}(r)$$

$$= \frac{3}{4}\int d\vec{r} \, g_{\vec{\sigma}_1\vec{\sigma}_2}(r). \tag{8.3.10}$$

For the free Fermi sea we have:

$$g_{\vec{\sigma}_1\vec{\sigma}_2}(r) = -\frac{l(k_F r)^2}{v}, \tag{8.3.11}$$

and recall from Prob 8.1

$$-\rho \int d^3r \frac{l(k_F r)^2}{v} = -1. \tag{8.3.12}$$

Therefore,

$$\frac{1}{N}\langle \phi_{FS}| \frac{1}{4}2\sum_{i<j}\vec{\sigma}_i \vec{\sigma}_j |\phi_{FS}\rangle = \frac{3}{4}\left(-\rho \int d^3r \frac{l(k_F r)^2}{v}\right) = -\frac{3}{4}. \tag{8.3.13}$$

On the other hand,

$$\frac{1}{N}\langle \phi_{FS}| \sum_i (\vec{\sigma}_i)^2 |\phi_{FS}\rangle = \frac{1}{N}\langle \phi_{FS}| \frac{3N}{4} |\phi_{FS}\rangle = \frac{3}{4}. \tag{8.3.14}$$

Finally,

$$\frac{1}{N}\langle \phi_{FS}| \vec{S}^2 |\phi_{FS}\rangle = \frac{3}{4} - \frac{3}{4} = 0. \tag{8.3.15}$$

Now, if we consider that $\vec{S}^2 = S_x^2 + S_y^2 + S_z^2$, where each contribution is positive defined, we can conclude that $|\phi_{FS}\rangle$ is an eigenstate of $\vec{S}^2$ with zero eigenvalue.

9 Dynamic structure functions

The dynamic structure function, closely tied to the linear response function, describes the system's reaction to external forces and encodes information about its excitations and correlations. It is thus an appropriate tool to study phenomena such as collective modes, particle-hole excitations, and energy dissipation. The linear response function quantifies the proportional relationship between an applied external perturbation and the induced response in the system, making it a cornerstone of many-body physics. Problems in this section also explore key concepts such as sum rules, which are integral constraints that provide insights into the conservation laws and spectral properties of the system.

Together, these topics equip readers with essential tools for analyzing the microscopic dynamics using the dynamic structure function as a starting point.

PROBLEM 9.1
Dynamic structure function for the Fermi sea at high momentum transfer

The dynamic structure function of the free Fermi sea can be expressed as

$$S(q,E) = \frac{v}{(2\pi)^3\rho} \int d\vec{k}\, \Theta(k_F - k)\, \Theta(|\vec{k}+\vec{q}| - k_F)\, \delta\left(E - \left(\frac{\hbar^2(\vec{k}+\vec{q})^2}{2m} - \frac{\hbar^2 k^2}{2m}\right)\right),$$

where q and E are the momentum and energy transferred to the system.

(a) Demonstrate that for $q > 2k_F$ the impulse approximation is exact.

(b) Calculate the dynamic structure function for $q > 2k_F$.

(a) For $q > 2k_F$ and $k < k_F$, the function $\Theta(|\vec{k}+\vec{q}| - k_F)$ is superfluous in the sense that for $q > 2k_F$ and $k < k_F$, $|\vec{k}+\vec{q}| > k_F$. Therefore, the dynamic structure function reduces to:

$$S(q,E) = \frac{v}{(2\pi)^3\rho} \int d\vec{k}\, \Theta(k_F - k)\, \delta\left(E - \left(\frac{\hbar^2(\vec{k}+\vec{q})^2}{2m} - \frac{\hbar^2 k^2}{2m}\right)\right)$$

$$= \frac{v}{(2\pi)^3\rho} \int d\vec{k}\, n(k)\, \delta\left(E - \left(\frac{\hbar^2(\vec{k}+\vec{q})^2}{2m} - \frac{\hbar^2 k^2}{2m}\right)\right),$$

where $n(k) = \Theta(k_F - k)$ is the momentum distribution. This last expression coincides with the impulse approximation.

(b) For $q > 2k_F$, we have

$$S(q,E) = \frac{v}{(2\pi)^3\rho} \int d\vec{k}\, \Theta(k_F - k)\, \delta\left(E - \left(\frac{\hbar^2(\vec{k}+\vec{q})^2}{2m} - \frac{\hbar^2 k^2}{2m}\right)\right).$$

Switching to spherical coordinates will allow us to ignore the Θ function and to compute the sum $(\vec{k}+\vec{q})^2$:

$$\int_{\mathbb{R}^3} d\vec{k} \longrightarrow 2\pi \int_0^{k_F} dk\, k^2 \int_0^{\pi} d\theta \sin\theta, \quad (\vec{k}+\vec{q})^2 = k^2 + q^2 + 2kq\cos\theta.$$

If we now use the change of variables $x = \cos\theta$ and basic properties of the Dirac δ function:

$$S(q,E) = \frac{v}{(2\pi)^2\rho} \int_0^{k_F} dk\, k^2 \int_{-1}^1 dx\, \delta\left(E - \left(\frac{2\hbar^2 kqx}{2m} + \frac{\hbar^2 q^2}{2m}\right)\right)$$

$$= \frac{v}{(2\pi)^2\rho} \int_0^{k_F} dk\, k^2 \int_{-1}^1 dx\, \frac{m}{\hbar^2 kq}\, \delta\left(\frac{Em}{\hbar^2 kq} - \frac{q}{2k} - x\right).$$

From the δ function we realize that we have contribution only if $x = Em/(\hbar^2 kq) - q/(2k)$ happens to be such that $-1 \le x \le 1$. Therefore we will have contribution different than zero, only if

$$-1 \le \frac{Em}{\hbar^2 kq} - \frac{q}{2k} \le 1,$$

which can be easily manipulated to get

$$-\frac{\hbar^2 kq}{m} + \frac{\hbar^2 q^2}{2m} \le E \le \frac{\hbar^2 kq}{m} + \frac{\hbar^2 q^2}{2m}.$$

For each q and E, even if E fulfills the previous condition, the integral over x imposes also a condition in the lower limit integration on k, as we require that $|x| < 1$. Therefore,

$$\left| \frac{Em}{\hbar^2 kq} - \frac{q}{2k} \right| < 1 \Rightarrow \left| \frac{1}{k} \left(\frac{Em}{\hbar^2 q} - \frac{q}{2} \right) \right| < 1 \Rightarrow k > \left| \frac{Em}{\hbar^2 q} - \frac{q}{2} \right|,$$

and we can write

$$S(q,E) = \frac{v}{(2\pi)^2 \rho} \int_{\frac{q}{2} \left| \frac{E}{\hbar^2 q^2} - 1 \right|}^{k_F} dk k^2 \frac{m}{\hbar^2 kq} = \frac{v}{(2\pi)^2 \rho} \frac{m}{\hbar^2 2q} k^2 \Big|_{\frac{q}{2} \left| \frac{E}{\hbar^2 q^2} - 1 \right|}^{k_F},$$

and finally

$$S(q,E) = \frac{v}{4\pi^2 \rho} \frac{m}{\hbar^2 q} \left[\frac{k_F^2}{2} - \frac{1}{2} \left(\frac{mE}{\hbar^2 q} - \frac{q}{2} \right)^2 \right].$$

PROBLEM 9.2
Sum rules of the dynamic structure function for $q \ge 2 k_F$

Use the definition of the dynamical structure function of the free Fermi sea for a momentum transferred $q \ge 2 k_F$:

$$S(q,E) = \frac{v}{(2\pi)^3 \rho} \int d\vec{k} \, \theta(k_F - k) \, \delta \left(E - \left(\frac{\hbar^2 (\vec{k} + \vec{q})^2}{2m} - \frac{\hbar^2 k^2}{2m} \right) \right),$$

where v is the spin-isospin degeneracy and k_F is the Fermi momentum, to

(a) Calculate the m_0 and m_1 energy-weighted sum rules.

(b) Calculate m_0 and m_1 using the explicit expression for $q \ge 2k_F$,

$$S(q,E) = \frac{v}{4\pi^2 \rho} \frac{m}{\hbar^2} \frac{1}{q} \left[\frac{k_F^2}{2} - \frac{1}{2} \left(\frac{mE}{\hbar^2 q} - \frac{q}{2} \right)^2 \right],$$

when $\hbar^2 q^2/2m - \hbar^2 k_F q/2m \leq E \leq \hbar^2 q^2/2m + \hbar^2 k_F q/2m$ and zero otherwise.

(a) The m_0 sum rule,

$$m_0(q) = \int_0^\infty dE \, S(q, E),$$

using the definition given above, we have:

$$
\int_0^\infty dE \, S(q, E) = \int_0^\infty dE \, \frac{V}{(2\pi)^3 \rho} \int d\vec{k} \, \theta(k_F - k) \\
\times \delta \left(E - \left(\frac{\hbar^2 (\vec{k} + \vec{q})^2}{2m} - \frac{\hbar^2 k^2}{2m} \right) \right).
$$

Changing the order of integration, i.e., integrating first over E, we can take advantage of the fact that:

$$
\int_0^\infty dE \, \delta \left(E - \left(\frac{\hbar^2 (\vec{k} + \vec{q})^2}{2m} - \frac{\hbar^2 k^2}{2m} \right) \right) = 1.
$$

Therefore,

$$
m_0(q) = \frac{V}{(2\pi)^3 \rho} \int d\vec{k} \, \theta(k_F - k) = \frac{V}{(2\pi)^3 \rho} \frac{4}{3} \pi k_F^3 = 1,
$$

as it should be, because

$$
\int_0^\infty dE \, S(q, E) = S(q),
$$

and for the free Fermi sea, $S(q) = 1$ when $q \geq 2k_F$.

On the other hand,

$$
m_1(q) = \int_0^\infty dE \, E \, S(q, E) = \int_0^\infty dE \, E \frac{V}{(2\pi)^3 \rho} \\
\times \int d\vec{k} \, \theta(k_F - k) \delta \left(E - \left(\frac{\hbar^2 (\vec{k} + \vec{q})^2}{2m} - \frac{\hbar^2 k^2}{2m} \right) \right).
$$

Again, one should perform first the integral over the energy:

$$
\int_0^\infty dE \, E \, \delta \left(E - \left(\frac{\hbar^2 (\vec{k} + \vec{q})^2}{2m} - \frac{\hbar^2 k^2}{2m} \right) \right) = \frac{\hbar^2 \vec{k} \cdot \vec{q}}{m} + \frac{\hbar^2 q^2}{2m}.
$$

Therefore,

$$m_1(q) = \frac{v}{(2\pi)^3\rho} \int d\vec{k}\, \theta(k_F - k)\left(\frac{\hbar^2\vec{k}\cdot\vec{q}}{m} + \frac{\hbar^2 q^2}{2m}\right).$$

Due to the angular integration, the term with $\hbar^2\vec{k}\cdot\vec{q}/m$ gives zero contribution. Then

$$m_1(q) = \frac{\hbar^2 q^2}{2m} \frac{v}{(2\pi)^3\rho} \int d\vec{k}\, \theta(k_F - k) = \frac{\hbar^2 q^2}{2m},$$

as it should be. Actually this result is true also for an interacting system, at any q.

(b) We now make the calculation using the explicit expression of the response for $q \geq 2k_F$. Using the limits for E obtained in Exercise 9.1:

$$m_0(q) = \int_{\frac{\hbar^2 q^2}{2m} - \frac{\hbar^2 k_F q}{m}}^{\frac{\hbar^2 q^2}{2m} + \frac{\hbar^2 k_F q}{m}} dE\, \frac{v}{4\pi^2\rho} \frac{m}{\hbar^2}\frac{1}{q}\left[\frac{k_F^2}{2} - \frac{1}{2}\left(\frac{mE}{\hbar^2 q} - \frac{q}{2}\right)^2\right].$$

Now the following change of variable is convenient: $E' = E - \hbar^2 q^2/2m$, then

$$m_0(q) = \int_{-\frac{\hbar^2 k_F q}{m}}^{\frac{\hbar^2 k_F q}{m}} dE'\, \frac{v}{4\pi^2\rho}\frac{m}{\hbar^2}\frac{1}{q}\left[\frac{k_F^2}{2} - \frac{1}{2}\left(\frac{mE}{\hbar^2 q} - \frac{q}{2}\right)^2\right]$$

$$= \frac{v}{4\pi^2\rho}\frac{m}{\hbar^2}\frac{1}{q}\int_{-\frac{\hbar^2 k_F q}{m}}^{\frac{\hbar^2 k_F q}{m}} dE'\left(\frac{k_F^2}{2} - \frac{1}{2}\frac{m^2}{\hbar^4 q^2}(E')^2\right)$$

$$= \frac{v}{4\pi^2\rho}\frac{m}{\hbar^2}\frac{1}{q}\left(\frac{k_F^2}{2}\frac{2\hbar^2 k_F q}{m} - \frac{1}{2}\frac{m^2}{\hbar^4 q^2}\frac{2}{3}\frac{\hbar^6 k_F^3 q^3}{m^3}\right)$$

$$= \frac{v}{4\pi^2\rho}\frac{m}{\hbar^2}\frac{1}{q}\frac{2}{3}k_F^3\frac{\hbar^2 q}{m} = 1.$$

Similarly for $m_1(q)$,

$$m_1(q) = \int_{\frac{\hbar^2 q^2}{2m} - \frac{\hbar^2 k_F q}{m}}^{\frac{\hbar^2 q^2}{2m} + \frac{\hbar^2 k_F q}{m}} dE\, E\, \frac{v}{4\pi^2\rho}\frac{m}{\hbar^2}\frac{1}{q}\left[\frac{k_F^2}{2} - \frac{1}{2}\left(\frac{mE}{\hbar^2 q} - \frac{q}{2}\right)^2\right],$$

and performing again the same change of variables, $E' = E - \hbar^2 q^2/2m$, we get:

$$m_1(q) = \frac{v}{4\pi^2\rho}\frac{m}{\hbar^2}\frac{1}{q}\int_{-\frac{\hbar^2 k_F q}{m}}^{\frac{\hbar^2 k_F q}{m}} dE'\left(E' + \frac{\hbar^2 q^2}{2m}\right)$$

$$\times \left(\frac{k_F^2}{2} - \frac{1}{2}\left(\frac{m}{\hbar^2 q}\left(E' + \frac{\hbar^2 q^2}{2m}\right) - \frac{q}{2}\right)^2\right),$$

and then

$$m_1(q) = \frac{v}{4\pi^2\rho}\frac{m}{\hbar^2}\frac{1}{q}\int_{-\frac{\hbar^2 k_F q}{m}}^{\frac{\hbar^2 k_F q}{m}} dE' \left(E' + \frac{\hbar^2 q^2}{2m}\right)\left(\frac{k_F^2}{2} - \frac{1}{2}\left(\frac{mE'}{\hbar^2 q}\right)^2\right).$$

The integral multiplying the term $\frac{\hbar^2 q^2}{2m}$ is 1, because it corresponds to the normalization integral calculated above for the m_0 sum rule, and the contribution of the terms with E' is zero, as it is an integral of an odd function over a symmetric interval with respect to the origin:

$$\int_{\frac{\hbar^2 k_F q}{m}}^{\frac{\hbar^2 k_F q m}{}} dE' E' \left(\frac{k_F^2}{2} - \frac{1}{2}\frac{m^2}{\hbar^4 q^2}(E')^2\right) = 0.$$

Therefore,

$$m_1(q) = \frac{\hbar^2 q^2}{2m}.$$

PROBLEM 9.3
Maximum and width of the dynamic structure function of the free Fermi sea for $q \geq 2\, k_F$

The dynamic structure function of the free Fermi sea for a momentum transferred $q \geq 2\, k_F$ is given by:

$$S(q,E) = \frac{v}{(2\pi)^3\rho}\int d\vec{k}\,\theta(k_F - k)\,\delta\left(E - \left(\frac{\hbar^2(\vec{k}+\vec{q})^2}{2m} - \frac{\hbar^2 k^2}{2m}\right)\right),$$

where v is the spin-isospin degeneracy and k_F is the Fermi momentum. Performing the angular integration:

$$S(q,E) = \frac{v}{(2\pi)^3\rho}2\pi\frac{m}{\hbar^2 q}\int_{\frac{q}{2}|\frac{E}{\frac{\hbar^2 q^2}{2m}}-1|}^{k_F} k\, dk,$$

we finally get:

$$S(q,E) = \frac{v}{4\pi^2\rho}\frac{m}{\hbar^2}\frac{1}{q}\left(\frac{k_F^2}{2} - \frac{1}{2}\left(\frac{mE}{\hbar^2 q} - \frac{q}{2}\right)^2\right)$$

when $\hbar^2 q^2/2m - \hbar^2 k_F q/2m \leq E \leq \hbar^2 q^2/2m + \hbar^2 k_F q/2m$ and zero otherwise.

(a) Find the value of E that maximizes $S(q,E)$ for a given q.

(b) Find the value of $S(q,E)$ at the maximum.

(c) Find the width of $S(q,E)$, i.e. the energy interval at which $S(q,E) \neq 0$.

(a) To find the maximum for a given q we can, by inspection of the definition of $S(q,E)$, see that $S(q,E)$ will be maximum when $E = \hbar^2 q^2/2m$, because in this case the lowest limit of the integral will be zero, and that will define the maximum value of the integral.

Another way to find the maximum is to take the analytic expression of $S(q,E)$ and find the E for which the derivative of $S(q,E)$ with respect to E is zero,

$$\frac{\partial S(q,E)}{\partial E} = 0 \implies 2\left(\frac{mE}{\hbar^2 q} - \frac{q}{2}\right) = 0 \implies E = \frac{\hbar^2 q^2}{2m}.$$

One can also check that the second derivative with respect to the energy is negative and therefore this stationary point corresponds to a maximum.

This maximum is known as the quasi-elastic peak, which means that all the energy associated to the momentum q has been transferred to a particle of momentum zero in the free Fermi sea as if the collision was elastic.

(b) The value at the maximum is given by:

$$S(q,E_{max}) = \frac{V}{4\pi^2 \rho} \frac{mk_F^2}{\hbar^2 2} \frac{1}{q} = \frac{3}{4} \frac{m}{\hbar^2 q k_F}.$$

Notice that $S(q,E)$ has dimensions of the inverse of energy and decreases inversely proportional to q.

(c) Taking the definition of $S(q,E)$ and looking where it is not zero, we immediately see that the width is given by

$$\frac{2\hbar^2 k_F q}{m}.$$

Notice that the width increases linearly with q, and the value at the maximum decreases inversely proportional to q, in such a way that the total surface below $S(q,E)$ for $q > 2k_F$ is equal to 1, independently of the value of q.

PROBLEM 9.4
Compton profile and Y–scaling for the free Fermi sea

The Compton profile of the free Fermi sea for a momentum transfer $q \geq 2\,k_F$ is given by:

$$J(Y) = 2\pi \int_{|Y|}^{\infty} dq\, q\, \bar{n}(q),$$

where $Y = mE/\hbar^2 q - q/2$ and

$$\bar{n}(q) = \frac{V}{(2\pi)^3 \rho} n(q),$$

with $n(q) = \theta(k_F - q)$. Then $\bar{n}(q)$ is normalized to unity:

$$\int d\vec{q}\, \bar{n}(q) = 1.$$

Actually, the definition of the Compton profile exploits the fact that $S(q,E)$ for large q ($q \geq 2k_F$ for the free Fermi sea) depends only on Y, i.e., scales on this variable. In fact,

$$J(Y) = \frac{m}{\hbar^2 q} S(q,E).$$

(a) Calculate $J(Y)$.

(b) Calculate m_0.

$$m_0 = \int_{-k_F}^{k_F} J(Y) dY.$$

(c) Calculate m_1 and m_2

$$m_1 = \int_{-k_F}^{k_F} Y J(Y), \quad m_2 = \int_{-k_F}^{k_F} Y^2 J(Y) dY = \frac{2m}{3\hbar^2} \langle \hat{t} \rangle,$$

where $\langle \hat{t} \rangle$ is the expectation value of the kinetic energy per particle in the free Fermi sea.

(a) The Compton profile is defined as

$$J(Y) = 2\pi \int_{|Y|}^{\infty} dq\, q\, \bar{n}(q).$$

Therefore,

$$J(Y) = 2\pi \frac{V}{(2\pi)^3 \rho} \int_{|Y|}^{\infty} \theta(k_F - q) q dq = 2\pi \frac{V}{(2\pi)^3 \rho} \int_{|Y|}^{k_F} q\, dq$$

$$= 2\pi \frac{V}{(2\pi)^3 \rho} \frac{1}{2} q^2 \Big|_{|Y|}^{k_F} = \frac{V}{(2\pi)^2 \rho} \frac{1}{2} (k_F^2 - Y^2),$$

which is a parabolic function of Y defined for $|Y| < k_F$. $J(Y)$ is an even function of Y, whose maximum is located at $Y = 0$, and whose value at the maximum is $J(0) = 3/(4k_F)$.

(b) We can calculate m_0 as:

$$m_0 = \int_{-k_F}^{k_F} J(Y)dY = \frac{v}{(2\pi)^2 \rho} \frac{1}{2} \int_{-k_F}^{k_F} (k_F^2 - Y^2)dY$$

$$= \frac{v}{(2\pi)^2 \rho} \frac{1}{2} \left[k_F^2 Y \Big|_{-k_F}^{k_F} - \frac{Y^3}{3} \Big|_{-k_F}^{k_F} \right] = \frac{v}{(2\pi)^2 \rho} \frac{1}{2} \left[2k_F^3 - \frac{2}{3} k_F^3 \right]$$

$$= \frac{v}{(2\pi)^3 \rho} \frac{4\pi}{3} k_F^3 = 1 \,.$$

(c) Let us calculate m_1:

$$m_1 = \int_{-k_F}^{k_F} YJ(Y)dY = 0$$

because it is the integral of an odd function over a symmetric interval with respect to the origin.

On the other hand, for m_2 we have:

$$m_2 = \int_{-\infty}^{\infty} Y^2 J(Y)dY = \frac{v}{(2\pi)^2 \rho} \frac{1}{2} \int_{-k_F}^{k_F} Y^2 (k_F^2 - Y^2)dY$$

$$= \frac{v}{(2\pi)^2 \rho} \frac{1}{2} \left[k_F^2 \frac{1}{3} Y^3 \Big|_{-k_F}^{k_F} - \frac{1}{5} Y^5 \Big|_{-k_F}^{k_F} \right] = \frac{v}{(2\pi)^2 \rho} \frac{1}{2} \left(\frac{2k_F^5}{3} - \frac{2k_F^5}{5} \right)$$

$$= \frac{v}{(2\pi)^3 \rho} \frac{4\pi k_F^3}{3} \frac{k_F^2}{5} = \frac{2m}{3\hbar^2} \frac{3}{5} \frac{\hbar^2 k_F^2}{2m} = \frac{2m}{3\hbar^2} \langle \hat{t} \rangle \,,$$

where

$$\langle \hat{t} \rangle = \frac{3}{5} \frac{\hbar^2 k_F^2}{2m}$$

is the average kinetic energy of the free Fermi sea.

10 Hartree-Fock

This chapter focuses on problems related to the Hartree-Fock method, an important method used in quantum many-body theory for many years. The Hartree-Fock approach approximates the many-electron wavefunction as a single Slater determinant, capturing the effects of exchange interactions between electrons while neglecting explicit electron correlation. The Hartree-Fock method also plays a foundational role in the development and application of Density Functional Theory, a self-consistent method that recasts the problem in terms of electron density, making it computationally efficient for large systems. By solving the equations iteratively, one can obtain self-consistent solutions for the electronic structure of atoms, molecules, and solids.

With these problems the reader will gain a thorough understanding of the core concepts, techniques, and limitations of this fundamental quantum mechanical approach.

PROBLEM 10.1
Hartree-Fock for nuclear matter with a simple Skyrme interaction

Consider uniform symmetric nuclear matter at a given density. The nucleon-nucleon interaction is described by means of an effective contact interaction

$$v(r_{ij}) = \left(t_0 + \frac{1}{6}t_3\rho^\alpha\right)\delta(\vec{r}_i - \vec{r}_j) \tag{10.1.1}$$

with $t_0 = -1794$ MeV fm^3, $t_3 = 12817$ MeV fm$^{3+\alpha}$ and $\alpha = 1/3$. Remember that for nucleons $\hbar^2/m = 41.4687$ MeV fm^2.

This density dependent effective interaction has been constructed to produce a good saturation density and binding energy.

(a) Calculate the diagonal two-body matrix elements taking plane waves normalized to volume as single-particle states.

(b) Calculate the energy in the Hartree-Fock approximation as a function of density for nuclear matter.

(c) Calculate the chemical potential and the pressure as a function of the density.

(d) Calculate the speed of sound in nuclear matter.

(a) The direct part of the diagonal two-body matrix element:

$$\left\langle \vec{k}_1 m_{s_1} m_{\tau_1} \vec{k}_2 m_{s_2} m_{\tau_2}\right|\left(t_0 + \frac{1}{6}t_3\rho^\alpha\right)\delta(\vec{r}_1 - \vec{r}_2)\left|\vec{k}_1 m_{s_1} m_{\tau_1} \vec{k}_2 m_{s_2} m_{\tau_2}\right\rangle$$

$$= \left(t_0 + \frac{1}{6}t_3\rho^\alpha\right)\frac{1}{\Omega^2}\int d\vec{r}_1 \int d\vec{r}_2 e^{-i\vec{k}_1\vec{r}_1}\, e^{-i\vec{k}_2\vec{r}_2}\delta(\vec{r}_1 - \vec{r}_2)\, e^{i\vec{k}_1\vec{r}_1}\, e^{i\vec{k}_2\vec{r}_2}$$

$$\langle m_{s_1} m_{s_2}|m_{s_1} m_{s_2}\rangle \langle m_{\tau_1} m_{\tau_2}|m_{\tau_1} m_{\tau_2}\rangle$$

$$= \frac{1}{\Omega^2}\left(t_0 + \frac{1}{6}t_3\rho^\alpha\right)\int d\vec{r}_1 = \frac{1}{\Omega}\left(t_0 + \frac{1}{6}t_3\rho^\alpha\right),$$

while the exchange term reads:

$$\left\langle \vec{k}_1 m_{s_1} m_{\tau_1} \vec{k}_2 m_{s_2} m_{\tau_2} \right| \left(t_0 + \frac{1}{6} t_3 \rho^\alpha \right) \delta(\vec{r}_1 - \vec{r}_2) \left| \vec{k}_2 m_{s_2} m_{\tau_2} \vec{k}_1 m_{s_1} m_{\tau_1} \right\rangle$$

$$= \left(t_0 + \frac{1}{6} t_3 \rho^\alpha \right) \frac{1}{\Omega^2} \int d\vec{r}_1 d\vec{r}_2 \, e^{-i\vec{k}_1 \vec{r}_1} e^{-i\vec{k}_2 \vec{r}_2} \delta(\vec{r}_1 - \vec{r}_2) e^{i\vec{k}_2 \vec{r}_1} e^{i\vec{k}_1 \vec{r}_2}$$

$$\langle m_{s_1} m_{s_2} | m_{s_2} m_{s_1} \rangle \langle m_{\tau_1} m_{\tau_2} | m_{\tau_2} m_{\tau_1} \rangle$$

$$= \left(t_0 + \frac{1}{6} t_3 \rho^\alpha \right) \frac{1}{\Omega^2} \delta_{m_{s_1} m_{s_2}} \delta_{m_{\tau_1} m_{\tau_2}} \int d\vec{r}_1 \, d\vec{r}_2 \, e^{i(\vec{k}_2 - \vec{k}_1)(\vec{r}_1 - \vec{r}_2)} \delta(\vec{r}_1 - \vec{r}_2)$$

$$= \frac{1}{\Omega} \left(t_0 + \frac{1}{6} t_3 \rho^\alpha \right) \delta_{m_{s_1} m_{s_2}} \delta_{m_{\tau_1} m_{\tau_2}} .$$

Finally the antisymmetrized two-body matrix element is:

$$\left\langle \vec{k}_1 m_{s_1} m_{\tau_1}, \vec{k}_2 m_{s_2} m_{\tau_2} \right| v(r_{12}) \left| \vec{k}_1 m_{s_1} m_{\tau_1}, \vec{k}_2 m_{s_2} m_{\tau_2} - \vec{k}_2 m_{s_2} m_{\tau_2}, \vec{k}_1 m_{s_1} m_{\tau_1} \right\rangle$$

$$= \frac{1}{\Omega} \left(t_0 + \frac{t_3}{6} \rho^\alpha \right) \left(1 - \delta_{m_{s_1} m_{s_2}} \delta_{m_{\tau_1} m_{\tau_2}} \right) .$$

(b) Now, to evaluate the expectation value of the interaction in the free Fermi sea we should sum up all the two-body matrix elements:

$$\frac{\langle \Phi_{\mathrm{FS}} | V | \Phi_{\mathrm{FS}} \rangle}{N} = \frac{1}{N} \frac{1}{2} \sum_{\substack{m_{s_1} m_{s_2} \\ m_{\tau_1} m_{\tau_2}}} \frac{\Omega}{(2\pi)^3} \frac{\Omega}{(2\pi)^3} \int_{|\vec{k}_1| \leq k_F} d\vec{k}_1 \int_{|\vec{k}_2| \leq k_F} d\vec{k}_2$$

$$\frac{1}{\Omega} \left(t_0 + \frac{t_3}{6} \rho^\alpha \right) \left(1 - \delta_{m_{s_1} m_{s_2}} \delta_{m_{\tau_1} m_{\tau_2}} \right) .$$

Performing the integrals and the summations we arrive to:

$$\frac{\langle \Phi_{\mathrm{FS}} | V | \Phi_{\mathrm{FS}} \rangle}{N} = \frac{1}{2} \frac{1}{N} \frac{\Omega}{(2\pi)^3} \frac{(t_0 + \frac{t_3}{6} \rho^\alpha)}{(2\pi)^3} \left(\frac{4}{3} \pi k_F^3 \right)^2 [\mathrm{Tr}(1) - \mathrm{Tr}(P_\sigma P_\tau)]$$

$$= \frac{1}{2} \frac{\Omega}{N} \frac{1}{(2\pi)^6} \left(\frac{4}{3} \pi k_F^3 \right)^2 \left(t_0 + \frac{t_3}{6} \rho^\alpha \right) \nu^2 \left(1 - \frac{1}{\nu} \right)$$

$$= \frac{1}{2} \rho \left(t_0 + \frac{t_3}{6} \rho^\alpha \right) \left(1 - \frac{1}{\nu} \right) = \frac{1}{2} \rho \left(t_0 + \frac{t_3}{6} \rho^\alpha \right) \frac{3}{4} ,$$

where ν is the spin-isospin degeneracy $\nu = 4$. Adding the average kinetic energy, we finally can write the energy per particle as

$$e(\rho) = \frac{\hbar^2}{2m} \frac{3}{5} \left(\frac{3\pi^2}{2} \right)^{2/3} \rho^{2/3} + \frac{1}{2} \rho \left(t_0 + \frac{t_3}{6} \rho^\alpha \right) \frac{3}{4} . \tag{10.1.2}$$

The energy has a minimum at the saturation density, $\rho_0 = 0.16 \; \mathrm{fm}^{-3}$ and $e_0 = -15.95 \; \mathrm{MeV}$. By definition, at ρ_0 the pressure is zero, and therefore the chemical

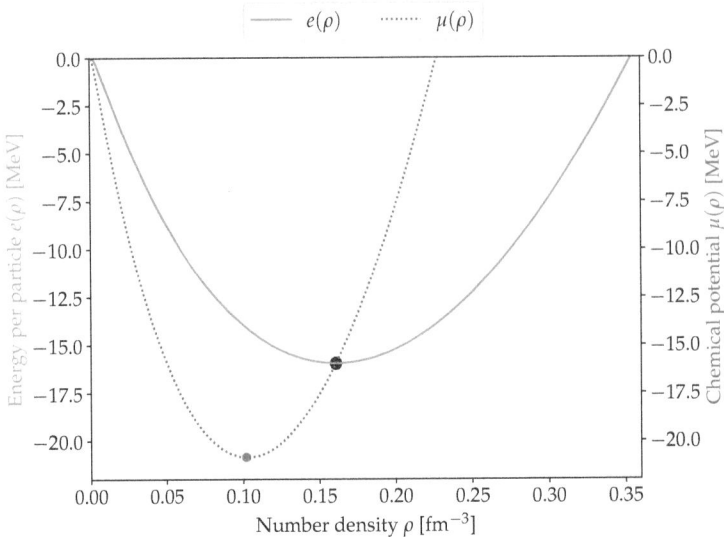

Figure 10.1 Energy per particle and chemical potential of nuclear matter as a function of the density. The black point marks the energy minimum at saturation density, and the gray one the spinodal point.

potential,

$$\mu(\rho) = e(\rho) + \frac{P(\rho)}{\rho}, \tag{10.1.3}$$

at ρ_0 coincides with the binding energy. Due to the contribution of the pressure, $\mu(\rho)$ grows faster, for $\rho > \rho_0$ than $e(\rho)$. On the other hand, for $\rho < \rho_0$, as $P(\rho) < 0$, $\mu(\rho)$ lies below $e(\rho)$. At $\rho = 0.354$ fm^{-3} the binding energy is zero.

(c) The chemical potential is

$$\mu(\rho) = \left(\frac{\partial E}{\partial N}\right)_\Omega = \left(\frac{\partial (Ne(\rho))}{\partial N}\right)_\Omega = e(\rho) + N\frac{\partial e}{\partial \rho}\left(\frac{\partial \rho}{\partial N}\right)_\Omega = e(\rho) + \rho\frac{\partial e(\rho)}{\partial \rho}, \tag{10.1.4}$$

and therefore

$$\mu(\rho) = \frac{\hbar^2}{2m}\frac{3}{5}\left(\frac{3\pi^2}{2}\right)^{2/3}\rho^{2/3} + \frac{1}{2}\rho\left(t_0 + \frac{t_3}{6}\rho^\alpha\right)\frac{3}{4}$$

$$+ \frac{\hbar^2}{2m}\frac{3}{5}\left(\frac{3\pi^2}{2}\right)^{2/3}\frac{2}{3}\rho^{2/3} + \frac{3}{8}\rho t_0 + \frac{1}{2}(\alpha+1)\rho^{\alpha+1}\frac{t_3}{6}\frac{3}{4}$$

$$= \frac{\hbar^2}{2m}\left(\frac{3\pi^2}{2}\right)^{2/3}\rho^{2/3} + \frac{3}{4}\rho t_0 + (\alpha+2)\rho^{\alpha+1}\frac{t_3}{6}\frac{3}{8},$$

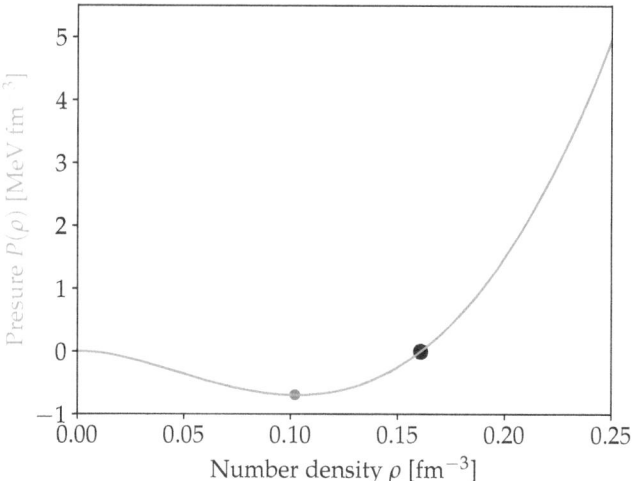

Figure 10.2 Pressure as a function of density. The black point marks the saturation density, and the grey one the spinodal point.

and the pressure is

$$P(\rho) = -\left(\frac{\partial E}{\partial \Omega}\right)_N = \rho^2 \frac{de(\rho)}{d\rho} = \rho[\mu(\rho) - e(\rho)]. \tag{10.1.5}$$

Therefore,

$$P(\rho) = \frac{\hbar^2}{2m} \frac{2}{5}\left(\frac{3\pi^2}{2}\right)^{2/3}\rho^{5/3} + \frac{3}{8}\rho^2 t_0 + \frac{3}{48}(\alpha+1)\rho^{\alpha+2}t_3. \tag{10.1.6}$$

The pressure is zero at the saturation density and has a minimum at the spinodal point, $\rho_s = 0.102$ fm^{-3}. Below this density, the system is mechanically unstable and the speed of sound is an imaginary number.

(d) The speed of sound can be calculated as

$$c_s(\rho) = \frac{1}{\hbar}\left(\frac{\hbar^2 c^2}{mc^2\rho}k^{-1}\right)^{1/2}, \tag{10.1.7}$$

where c is the speed of light and k^{-1} is the incompressibility

$$k^{-1}(\rho) = \rho\left(\frac{dP(\rho)}{d\rho}\right), \tag{10.1.8}$$

and therefore,

$$k^{-1}(\rho) = \frac{\hbar^2}{2m}\frac{2}{3}\left(\frac{3\pi^2}{2}\right)^{2/3}\rho^{5/3} + \frac{6}{8}\rho^2 t_0 + \frac{3(\alpha+1)(\alpha+2)}{48}\rho^{\alpha+2}t_3. \tag{10.1.9}$$

Finally, the speed of sound in units of the speed of light is:

$$\frac{c_s}{c} = \left(\frac{k^{-1}}{mc^2\rho}\right)^{1/2}.$$ (10.1.10)

PROBLEM 10.2
Hartree-Fock for a system of N fermions enclosed in a one-dimensional box

Consider a system of N fermions of mass m and spin $1/2$, all with the same spin orientation, in a one-dimensional box. The single-particle Hamiltonian of the system is

$$\hat{H}_0 = \sum_{i=1}^{N} \hat{h}_0(i) = \sum_{i=1}^{n} -\frac{\hbar^2}{2m}\nabla_i^2 + \sum_{i=1}^{N} U(x_i),$$ (10.2.1)

where $U(x)$ is a square well potential with infinite walls: $U(x) = 0$ if $0 < x < a$ and $U(x) = \infty$ if $x > a$ or $x < 0$. The single-particle eigenfunctions and their eigenenergies are:

$$\phi_k = \left(\frac{2}{a}\right)^{1/2}\sin\left(\frac{\pi kx}{a}\right), \quad \varepsilon_k = \frac{\hbar^2\pi^2k^2}{2ma^2}, \quad k = 1,2,3,....$$ (10.2.2)

Suppose that there is a two-body interaction of the type $\hat{V} = \sum_{i<j} v(x_i, x_j)$ with $v(x_i, x_j) = C$, where C is a constant measured in units of energy. Then the total Hamiltonian is $\hat{H} = \hat{H}_0 + \hat{V}$.

(a) Calculate the two-body matrix element

$$\langle ij| v |kl - lk\rangle$$

where $|kl - lk\rangle = |kl\rangle - |lk\rangle$.

(b) For this Hamiltonian, the single-particle wavefunctions solution of the Hartree-Fock equations are the single-particle wavefunctions defined above. Calculate the single-particle Hartree-Fock energies:

$$\varepsilon(i) = \langle i|\hat{h}_0|i\rangle + U_{HF}(i),$$ (10.2.3)

$$U_{HF}(i) = \sum_{j=1}^{N} \langle ij| v |ij - ji\rangle .$$ (10.2.4)

(c) Calculate the total energy by using

$$E_{HF} = \sum_{i=1}^{N} \varepsilon_{HF}(i) - \frac{1}{2}\sum_{i=1}^{N} U_{HF}(i).$$ (10.2.5)

(d) Check that this energy coincides with the expectation value , $\langle\Psi_{\mathrm{HF}}|\hat{H}|\Psi_{\mathrm{HF}}\rangle$, where $|\Psi_{\mathrm{HF}}\rangle$ is the Slater determinant built with the lowest N single-particle wavefunctions.

(e) Check that $\hat{H}$ in second quantization is written as

$$\hat{H} = \sum_i \varepsilon_i \hat{a}_i^\dagger \hat{a}_i + \frac{1}{2}C\hat{N}^2 - \frac{1}{2}C\hat{N}, \qquad (10.2.6)$$

where $\hat{a}_i^\dagger (\hat{a}_i)$ are the creation (annihilation) operators associated to the single-particle basis $\phi_i(x)$, and $\hat{N} = \sum_i \hat{a}_i^\dagger \hat{a}_i$ is the number of particles operator.

(f) Demonstrate that the state $|\Psi\rangle = \left(\prod_{i=1}^N \hat{a}_i^\dagger\right)|0\rangle$ is an eigenstate of $\hat{H}$.

(a) Taking into account the orthonormality of the single-particle states and the fact that the interaction is a constant, we have

$$\langle ij|v|kl - lk\rangle = C\langle ij|kl - lk\rangle = C(\delta_{ik}\delta_{jl} - \delta_{il}\delta_{jk}). \qquad (10.2.7)$$

(b) The HF single-particle energies are given by

$$\varepsilon_{\mathrm{HF}}(i) = \langle i|\hat{h}_o|i\rangle + U_{\mathrm{HF}}(i), \qquad (10.2.8)$$

where

$$\langle i|\hat{h}_o|i\rangle = \frac{\hbar^2 \pi^2 i^2}{2ma^2} \qquad (10.2.9)$$

and

$$U_{\mathrm{HF}}(i) = \sum_{j\leq N} \langle ij|v|ij - ji\rangle = C\sum_{j\leq N}(1 - \delta_{ij}) = C(N-1), \qquad (10.2.10)$$

which is a reasonable result, as $U_{\mathrm{HF}}(i)$ takes into account the interaction of the state i with the rest of occupied states.

(c) The total HF energy is:

$$E_{\mathrm{HF}} = \sum_{i=1}^N \varepsilon_{\mathrm{HF}}(i) - \frac{1}{2}\sum_{i=1}^N U_{\mathrm{HF}}(i). \qquad (10.2.11)$$

Taking into account all the previous results we have:

$$
\begin{aligned}
E_{HF} &= \frac{\hbar^2 \pi^2}{2ma^2} \sum_{i=1}^{N} i^2 + \sum_{i=1}^{N} C(N-1) - \frac{1}{2} \sum_{i=1}^{N} C(N-1) \\
&= \frac{\hbar^2 \pi^2}{2ma^2} \sum_{i=1}^{N} i^2 + CN(N-1) - \frac{1}{2} CN(N-1) \\
&= \frac{\hbar^2 \pi^2}{2ma^2} \sum_{i=1}^{N} i^2 + \frac{1}{2} CN(N-1).
\end{aligned}
$$

(d) The Hartree-Fock energy can be also calculated as the expectation value of $\hat{H}$ in the HF wavefunction

$$
E_{HF} = \langle \Psi_{HF} | \hat{H} | \Psi_{HF} \rangle = \sum_{i=1}^{N} \frac{\hbar^2 \pi^2}{2ma^2} i^2 + \sum_{i<j}^{N} \langle ij|v|ij - ji \rangle , \qquad (10.2.12)
$$

which corresponds to the evaluation of a one-body operator $\hat{H}_0$ and a two-body operator $\hat{V}$, in the Slater determinant built with the first N single-particle solutions of the HF equations. Then

$$
E_{HF} = \sum_{i=1}^{N} \frac{\hbar^2 \pi^2}{2ma^2} i^2 + C \frac{N(N-1)}{2}. \qquad (10.2.13)
$$

(e) The Hamiltonian in the second quantization for this specific case, in which the interaction is a constant, is written as

$$
\hat{H} = \varepsilon_i \hat{a}_i^\dagger \hat{a}_i + \frac{1}{2} \sum_{ij} \langle ij|v|ij \rangle \hat{a}_i^\dagger \hat{a}_j^\dagger \hat{a}_j \hat{a}_i , \qquad (10.2.14)
$$

where

$$
\varepsilon_i = \frac{\hbar^2 \pi^2 i^2}{2ma^2} \quad , \quad \langle ij|v|ij \rangle = C. \qquad (10.2.15)
$$

On the other hand,

$$
\hat{a}_i^\dagger \hat{a}_j^\dagger \hat{a}_j \hat{a}_i = -\hat{a}_i^\dagger \hat{a}_i \delta_{ij} + \hat{a}_i^\dagger \hat{a}_i \hat{a}_j^\dagger \hat{a}_j . \qquad (10.2.16)
$$

Therefore,

$$
\begin{aligned}
\frac{1}{2} \sum_{ij} \langle ij|v|ij \rangle \hat{a}_i^\dagger \hat{a}_j^\dagger \hat{a}_j \hat{a}_i &= \frac{1}{2} C \sum_{ij} (-\hat{a}_i^\dagger \hat{a}_i \delta_{ij} + \hat{a}_i^\dagger \hat{a}_i \hat{a}_j^\dagger \hat{a}_j) \\
&= -\frac{1}{2} C \sum_i \hat{a}_i^\dagger \hat{a}_i + \frac{1}{2} C (\sum_i \hat{a}_i^\dagger \hat{a}_i)(\sum_j \hat{a}_j^\dagger \hat{a}_j) \\
&= -\frac{1}{2} C \hat{N} + \frac{1}{2} C \hat{N}^2 .
\end{aligned}
$$

Therefore

$$
\hat{H} = \sum_i \varepsilon_i \hat{a}_i^\dagger \hat{a}_i + \frac{1}{2} \hat{N}^2 - \frac{1}{2} C \hat{N}. \qquad (10.2.17)
$$

(f) In this case, for a constant interaction, the HF-ground state is an eigenstate. In fact,

$$\left(\sum_i \varepsilon_i \hat{a}_i^\dagger \hat{a}_i\right)\left(\prod_{j=1}^N \hat{a}_j^\dagger\right)|0\rangle = \left(\sum_{i=1}^N \varepsilon_i\right)\left(\prod_{j=1}^N \hat{a}_j^\dagger\right)|0\rangle = \left(\sum_{i=1}^N \varepsilon_i\right)|\Psi_{\mathrm{HF}}\rangle,$$

$$(10.2.18)$$

and the interaction term is

$$\left(\frac{1}{2}C\hat{N}^2 - \frac{1}{2}C\hat{N}\right)|\Psi_{\mathrm{HF}}\rangle = \frac{1}{2}C(N^2 - N)|\Psi_{\mathrm{HF}}\rangle = \frac{1}{2}CN(N-1)|\Psi_{\mathrm{HF}}\rangle.$$

$$(10.2.19)$$

Therefore, we can conclude that

$$\hat{H}|\Psi_{\mathrm{HF}}\rangle = \left(\frac{\hbar^2\pi^2}{2ma^2}\sum_{i=1}^N i^2 + \frac{1}{2}CN(N-1)\right)|\Psi_{\mathrm{HF}}\rangle.$$

$$(10.2.20)$$

PROBLEM 10.3
Single-particle potential for a uniform system of fermions in the Hartree-Fock approximation

Consider uniform system of fermions of spin $1/2$, with the same number of fermions with spin up and spin down, interacting through an interaction depending only on the distance:

$$V = \sum_{i<j} v(r_{ij}).$$

$$(10.3.1)$$

(a) Show that the single-particle potential can be written in the following way:

$$U(k) = \rho \int d\vec{r}\, v(r)\left(1 - \frac{l(k_{\mathrm{F}}r)}{\nu}e^{i\vec{k}\vec{r}}\right),$$

$$(10.3.2)$$

where $l(k_{\mathrm{F}}r)$ is the so-called Slater function and ν is the spin degeneracy.

(b) Demonstrate that the expectation value $\langle\Phi_{\mathrm{FS}}|V|\Phi_{\mathrm{FS}}\rangle$ can be calculated as

$$\frac{\langle\Phi_{\mathrm{FS}}|V|\Phi_{\mathrm{FS}}\rangle}{N} = \frac{1}{2}\frac{\nu}{\rho}\frac{1}{(2\pi)^3}\int_{|\vec{k}|\leq k_{\mathrm{F}}} d\vec{k}\, U(k).$$

$$(10.3.3)$$

(a) The single-particle potential for k is the sum of the interactions of the particle with momentum $\vec{k}$ with all the other particles. For a uniform system and an isotropic interaction, the potential depends only on the modulus of $\vec{k}$.

$$U(k) = \frac{1}{V} \sum_{k m_\sigma' m_\sigma} \langle \vec{k} m_\sigma \vec{k}' m_\sigma' | v(r_{12}) | \vec{k} m_\sigma \vec{k}' m_\sigma' - \vec{k}' m_\sigma' \vec{k} m_\sigma \rangle$$

$$= \frac{1}{V} \frac{\Omega}{(2\pi)^3} \int_{|\vec{k}'| \leq k_F} d\vec{k}' \sum_{m_\sigma' m_\sigma} \frac{1}{\Omega^2} \int d\vec{r}_1 d\vec{r}_2\, v(r_{12})$$

$$\left[e^{-i\vec{k}\vec{r}_1} e^{-i\vec{k}' r_2} \langle m_\sigma m_\sigma' | \left(e^{i\vec{k}\vec{r}_1} e^{i\vec{k}' r_2} | m_\sigma m_\sigma' \rangle - e^{i\vec{k}'\vec{r}_1} e^{i\vec{k} r_2} | m_\sigma' m_\sigma \rangle \right) \right]$$

$$= \frac{1}{V} \frac{\Omega}{(2\pi)^3} \frac{1}{\Omega^2} \int_{|\vec{k}'| \leq k_F} d\vec{k}' \int d\vec{r}_1 d\vec{r}_2\, v(r_{12})$$

$$\left(\mathrm{Tr}(I) - \mathrm{Tr}(P_\sigma)\, e^{i\vec{k}(\vec{r}_2 - \vec{r}_1)} e^{i\vec{k}'(\vec{r}_1 - \vec{r}_2)} \right).$$

Now, one should take into account the following integrals:

$$\int_{|\vec{k}| \leq k_F} d\vec{k} = \frac{4}{3} \pi k_F^3 = (2\pi)^3 \frac{\rho}{V} \tag{10.3.4}$$

and

$$\int_{|\vec{k}| \leq k_F} d\vec{k}\, e^{i\vec{k}\vec{r}} = \frac{(2\pi)^3 \rho}{V} l(k_F r). \tag{10.3.5}$$

Taking into account these integrals, one can integrate first over momenta:

$$U(k) = \frac{1}{V} \frac{\Omega}{(2\pi)^3} \frac{v^2}{\Omega^2} \int d\vec{r}_1 d\vec{r}_2\, v(r_{12}) \int_{|\vec{k}'| \leq k_F} d\vec{k}' \left(1 - \frac{1}{v} e^{i\vec{k}'(\vec{r}_1 - \vec{r}_2)} e^{i\vec{k}(\vec{r}_2 - \vec{r}_1)} \right),$$
$$\tag{10.3.6}$$

obtaining

$$U(k) = \frac{1}{V} \frac{\Omega}{(2\pi)^3} \frac{1}{\Omega^2} v^2 \int d\vec{r}_1 d\vec{r}_2\, v(r_{12}) \left(\frac{(2\pi)^3 \rho}{v} - \frac{(2\pi)^3 \rho}{v^2} l(k_F r_{12}) e^{i\vec{k}(\vec{r}_2 - \vec{r}_1)} \right) \tag{10.3.7}$$

and finally

$$U(k) = \rho \int d\vec{r}\, v(r) \left(1 - \frac{l(k_F r)}{v} e^{i\vec{k}\vec{r}} \right) \tag{10.3.8}$$

that will depend only on the modulus of $\vec{k}$.

(b) Now, integrating the single-particle potential we recover the expectation value of the interaction. In fact,

$$\frac{\langle \Phi_{FS} | V | \Phi_{FS} \rangle}{N} = \frac{1}{2} \frac{\Omega}{(2\pi)^3 N} V \int_{|\vec{k}| \leq k_F} d\vec{k}\, U(k)$$

$$= \frac{1}{2} \frac{1}{(2\pi)^3} V \frac{1}{\rho} \int_{|\vec{k}| \leq k_F} d\vec{k}\, \rho \int d\vec{r}\, v(r) \left(1 - \frac{l(k_F r)}{v} e^{i\vec{k}\vec{r}} \right),$$

changing the order of integration:

$$\frac{\langle \Phi_{FS}|V|\Phi_{FS} \rangle}{N} = \frac{1}{2}\frac{V}{(2\pi)^3} \int d\vec{r}\, v(r) \int_{|\vec{k}| \leq k_F} d\vec{k} \left(1 - \frac{l(k_F r)}{V} e^{i\vec{k}\vec{r}} \right)$$

$$= \frac{1}{2}\frac{V}{(2\pi)^3} \int d\vec{r}\, v(r) \left(\frac{(2\pi)^3 \rho}{V} - \frac{l(k_F r)}{V} l(k_F r)\frac{(2\pi)^3 \rho}{V} \right)$$

$$= \frac{1}{2}\rho \int d\vec{r}\, v(r) \left(1 - \frac{l(k_F r)^2}{V} \right) = \frac{1}{2}\rho \int d\vec{r}\, v(r) g_{FS}(r),$$

where $g_{FS}(r)$ is the two-body distribution function associated to the free Fermi sea (see Problem 8.2).

11 Density matrices

This chapter delves into the concept of density matrices, a fundamental tool in quantum mechanics for describing the state of a many-body system. Density matrices provide a compact and powerful representation of quantum states, encompassing both pure and mixed states. The study of one-body and two-body density matrices is particularly important for understanding the behavior of many-body systems, especially in contexts like quantum gases, solid-state physics, and quantum chemistry. These matrices encapsulate information about particle distributions and correlations, enabling the derivation of key properties such as energy, momentum, and entropy.

By working through the problems in this chapter, readers will gain a deeper understanding of how density matrices connect the microscopic wavefunction to experimentally accessible quantities, providing a versatile framework for studying non-interacting and interacting systems alike.

PROBLEM 11.1
Derivatives of the one-body density matrix at $r = 0$.

One way to calculate the one-body density matrix of a homogeneous system is through the momentum distribution of the system:

$$\rho(r) = \frac{V}{(2\pi)^3} \int d\vec{k}\, n(k)\, e^{i\vec{k}\cdot\vec{r}},$$

where $n(k)$ is the momentum distribution, independent of the spin orientation, and normalized such that

$$\rho_0 = \frac{V}{(2\pi)^3} \int d\vec{k}\, n(k),$$

where ρ_0 is the density of the homogeneous system. Notice that the diagonal of the density matrix fulfills that $\rho(r = 0) = \rho_0$.

(a) Do the series expansion of $\rho(r)$ around $r = 0$ and show that all odd derivatives of $\rho(r)$ at the origin are zero, and that

$$\frac{d^2\rho(r)}{dr^2} = -\frac{V}{(2\pi)^3}\frac{1}{3}\int d\vec{k}\, k^2\, n(k).$$

(b) Show that the kinetic energy per particle can be related to the second derivative of the one-body density matrix at the origin:

$$\frac{\langle \hat{T} \rangle}{N} = -\frac{3}{2}\frac{\hbar^2}{m}\frac{1}{\rho_0}\frac{d^2\rho(r)}{dr^2}\bigg|_{r=0}.$$

(c) Consider a free Fermi sea of fermions of spin 1/2, whose one-body density matrix is given by $\rho(r) = \rho_0\, l(k_F r)$, where k_F is the Fermi momentum and $l(k_F r)$ is the Slater function , defined as

$$l(z) = \frac{3 j_1(z)}{z},$$

where $j_1(z)$ is the spherical Bessel function of order one:

$$j_1(z) = \frac{\sin(z) - z\cos(z)}{z^2}.$$

Do the series expansion of $\rho(r)$ around $r = 0$ and check the previous properties for a free Fermi sea with spin degeneracy 2 and $n(k) = \theta(k_F - k)$.

(a) To perform the series expansion of $\rho(r)$ we start from its definition, as the Fourier transform of the momentum distribution. Notice that here we consider that the system is isotropic and therefore $n(k)$ depends only on the modulus of $\vec{k}$. On the other hand both spin orientations, up and down with respect to the spin quantization axis, have the same $n(k)$. Then $\rho(r)$ is given by

$$\rho(r) = \frac{v}{(2\pi)^3} \int d\vec{k}\, n(k)\, e^{i\vec{k}\cdot\vec{r}}.$$

Now we perform the angular integrations and check that actually $\rho(r)$ depends only on the modulus of $\vec{r}$:

$$
\begin{aligned}
\rho(r) &= \frac{v}{(2\pi)^3} \int_0^\infty dk\, k^2 \int_0^{2\pi} d\phi \int_{-1}^1 dx\, e^{-i\vec{k}\cdot\vec{r}} n(k) \\
&= \frac{v}{(2\pi)^3} 2\pi \int_0^\infty dk\, k^2 n(k) \int_{-1}^1 dx\, (\cos(krx) + i\sin(krx)) \\
&= \frac{v}{(2\pi)^3} 2\pi \int_0^\infty dk\, k^2 n(k) \left. \frac{\sin(krx)}{kr} \right|_{-1}^1 \\
&= \frac{v}{(2\pi)^3} 4\pi \int dk\, k^2\, n(k) \frac{\sin(kr)}{kr}.
\end{aligned}
$$

Now we take into account the Taylor expansion of $\sin(kr)/kr$,

$$\frac{\sin(kr)}{kr} \simeq 1 - \frac{(kr)^2}{3!} + \frac{(kr)^4}{5!} - \cdots.$$

Therefore,

$$
\rho(r) = \frac{V}{(2\pi)^3} 4\pi \int dk\, k^2 n(k) \left[1 - \frac{(kr)^2}{3!} + \frac{(kr)^4}{5!} - \cdots \right]
$$

$$
= \frac{V}{(2\pi)^3} 4\pi \left[\int dk\, k^2 n(k) - \left(\int dk\, k^4 n(k) \right) \frac{r^2}{3!} + \left(\int dk\, k^6 n(k) \right) \frac{r^4}{5!} + \cdots \right]
$$

$$
= \frac{V}{(2\pi)^3} \int d\vec{k}\, n(k) - \left[\frac{V}{(2\pi)^3} \int d\vec{k}\, k^2 n(k) \right] \frac{r^2}{3!} + \left[\frac{V}{(2\pi)^3} \int d\vec{k}\, k^6 n(k) \right] \frac{r^4}{5!}
$$

$$
+ \cdots .
$$

Then, we should compare the previous expansion with the Taylor expansion of $\rho(r)$:

$$
\rho(r) = \rho(0) + r \left. \frac{d\rho}{dr} \right|_{r=0} + \frac{1}{2!} \left. \frac{d^2\rho}{dr^2} \right|_{r=0} r^2 + \frac{1}{3!} \left. \frac{d^3\rho}{dr^3} \right|_{r=0} r^3 + \cdots ,
$$

and by equating the coefficients we can identify the derivatives of $\rho(r)$. The first thing to realize is that all odd derivatives of $\rho(r)$ for $r = 0$ are zero. The second thing we can conclude is :

$$
\rho(0) = \frac{V}{(2\pi)^3} 4\pi \int dk\, k^2 n(k) = \rho_0 ,
$$

and for the second derivative:

$$
\left. \frac{1}{2} \frac{d^2\rho}{dr^2} \right|_{r=0} = -\frac{1}{3!} \left[\frac{V}{(2\pi)^3} \int d\vec{k}\, k^2 n(k) \right] .
$$

Therefore,

$$
\left. \frac{d^2\rho}{dr^2} \right|_{r=0} = -\frac{1}{3} \frac{V}{(2\pi)^3} \int d\vec{k}\, k^2 n(k) .
$$

(b) Then this second derivative could be related with the kinetic energy per particle:

$$
\frac{\langle \hat{T} \rangle}{N} = \frac{V}{(2\pi)^3} \frac{1}{\rho_0} \int d\vec{k}\, \frac{\hbar^2 k^2}{2m} n(k) = -\frac{3}{2} \frac{\hbar^2}{m} \frac{1}{\rho_0} \left. \frac{d^2\rho}{dr^2} \right|_{r=0} ,
$$

which is valid for any homogeneous Fermi system.

(c) In our particular case, i.e., the Fermi sea, we have

$$
\rho(r) = \rho_0\, l(k_F r) .
$$

The Slater function, $l(k_F r)$, can be developed in powers of $k_F r$ by using the expansion of the spherical Bessel function $j_1(z)$,

$$
j_1(z) = \frac{z}{3} \left(1 - \frac{\frac{1}{2} z^2}{15} + \frac{\frac{1}{4} z^4}{2\,5\,7} - \cdots \right) = \frac{z}{3} - \frac{z^3}{30} + \frac{z^5}{840} - \cdots .
$$

Doing this we obtain:

$$\rho(r) = \rho_0 \left[1 - \frac{k_F^2}{10} r^2 + \frac{k_F^4}{280} r^4 - \cdots \right],$$

and we can immediately identify

$$\rho(0) = \rho_0, \qquad \left. \frac{d\rho(r)}{dr} \right|_{r=0} = 0, \qquad \left. \frac{d^2\rho}{d^2 r} \right|_{r=0} = -\rho_0 \frac{k_F^2}{5}.$$

Therefore,

$$\frac{\langle \hat{T} \rangle}{N} = \frac{\nu}{(2\pi)^3} \frac{1}{\rho_0} \int d\vec{k} \frac{\hbar^2 k^2}{2m} n(k) = -\frac{3}{2} \frac{\hbar^2}{m} \frac{1}{\rho_0} \left. \frac{d^2\rho}{dr^2} \right|_{r=0} = \frac{3}{5} \frac{\hbar^2 k_F^2}{2m},$$

which is nothing other than the kinetic energy per particle of the free Fermi sea.

PROBLEM 11.2
Sequential condition of the three-body distribution function.

The three-body distribution function of a many-body system is defined as,

$$g_3(\vec{r}_1, \vec{r}_2, \vec{r}_3) = \frac{N(N-1)(N-2)}{\rho^3} \int d\vec{r}_4 \dots d\vec{r}_N \, \Psi^*(\vec{r}_1, \dots, \vec{r}_N) \Psi(\vec{r}_1, \dots, \vec{r}_N),$$

where $\Psi(\vec{r}_1, \dots, \vec{r}_N)$ is a unit-normalized pure state. For an infinite, homogeneous system, g_3 depends only on the distances between the three particles r_{12}, r_{13}, r_{23}. Usually one writes $g_3(\vec{r}_{12}, \vec{r}_{13})$.

In the case of the free Fermi sea we have,

$$g_3(\vec{r}_{12}, \vec{r}_{13}) = 1 - \frac{1}{\nu} l^2(k_F r_{12}) - \frac{1}{\nu} l^2(k_F r_{13}) - \frac{1}{\nu} l^2(k_F r_{23})$$
$$+ \frac{2}{\nu} l(k_F r_{12}) l(k_F r_{13}) l(k_F r_{23}),$$

where ν is the spin degeneracy.

(a) Derive the sequential condition

$$\rho \int d\vec{r}_3 \, g_3(\vec{r}_{12}, \vec{r}_{13}) = (N-2) \, g_2(r_{12}),$$

where $g_2(r_{12})$ is the two-body distribution function .

(b) Check the validity of this relation for the free Fermi sea.

(a) To derive the sequential condition we should use the definition of the $n-$body distribution functions. Actually this property is valid both for Bose and Fermi systems.

$$\rho \int g_3(\vec{r}_{12}, \vec{r}_{13}) d\vec{r}_3 = \rho \frac{N(N-1)(N-2)}{\rho^3} \int d\vec{r}_3 d\vec{r}_4 \ldots d\vec{r}_N \ \Psi^*(\vec{r}_1, \ldots, \vec{r}_N) \Psi(\vec{r}_1, \ldots, \vec{r}_N)$$

$$= (N-2) \frac{N(N-1)}{\rho^2} \int d\vec{r}_3 d\vec{r}_4 \ldots d\vec{r}_N \ \Psi^*(\vec{r}_1, \ldots, \vec{r}_N) \Psi(\vec{r}_1, \ldots, \vec{r}_N)$$

$$= (N-2) g_2(\vec{r}_1, \vec{r}_2) = (N-2) g_2(r_{12}).$$

(b) To check the validity of the sequential condition for the $g_3(\vec{r}_{12}, \vec{r}_{13})$ of the free Fermi sea we need several properties of the Slater functions,

$$\int d\vec{r} \ l^2(k_F r) = \frac{v}{\rho},$$

that was derived in Problem 8.1.

We still need another relation,

$$\rho \int d\vec{r}_3 \ l(k_F r_{13}) \ l(k_F r_{23}) = \rho \int d\vec{r}_3 \ l(k_F r_{13}) \ l(k_F r_{32}) = v \ l(k_F r_{12}).$$

To derive this relation one should use the explicit definition of the Slater function,

$$\rho \int d\vec{r}_3 \ l(k_F r_{13}) \ l(k_F r_{32}) = \rho \int d\vec{r}_3 \frac{v}{(2\pi)^3 \rho} \int_{|\vec{k}| \leq k_F} d\vec{k} \ e^{i\vec{k} \cdot (\vec{r}_3 - \vec{r}_1)}$$

$$\times \frac{v}{(2\pi)^3 \rho} \int_{|\vec{k}'| \leq k_F} d\vec{k}' \ e^{i\vec{k}' \cdot (\vec{r}_2 - \vec{r}_3)}.$$

Now we change the order of integration,

$$\rho \int d\vec{r}_3 l(k_F r_{13}) l(k_F r_{32}) = \rho \int_{|\vec{k}| \leq k_F} d\vec{k} \int_{|\vec{k}'| \leq k_F} d\vec{k}' \ \frac{(v)^2}{\rho^2 (2\pi)^3} e^{-i\vec{k} \cdot \vec{r}_1} e^{i\vec{k}' \cdot \vec{r}_2}$$

$$\times \frac{1}{(2\pi)^3} \int d\vec{r}_3 e^{i\vec{r}_3 \cdot (\vec{k} - \vec{k}')}$$

$$= \rho \int_{|\vec{k}| \leq k_F} d\vec{k} \int_{|\vec{k}'| \leq k_F} d\vec{k}' \ \frac{(v)^2}{\rho^2 (2\pi)^3} e^{-i\vec{k} \cdot \vec{r}_1} e^{i\vec{k}' \cdot \vec{r}_2} \delta(\vec{k} - \vec{k}')$$

$$= \rho \int_{|\vec{k}| \leq k_F} \frac{(v)^2}{\rho^2 (2\pi)^3} e^{i\vec{k} \cdot (\vec{r}_2 - \vec{r}_1)} = v \ l(k_F r_{12}).$$

Now we can do the explicit calculation for the three-body distribution function of the free Fermi sea $g_3^{FS}(\vec{r}_{12}, \vec{r}_{13})$,

$$\rho \int d\vec{r}_3 \left[1 - \frac{1}{v}l^2(k_F r_{12}) - \frac{1}{v}l^2(k_F r_{13}) - \frac{1}{v}l^2(k_F r_{23}) \right.$$

$$\left. + \frac{2}{v}l(k_F r_{12})l(k_F r_{13})l(k_F r_{23}) \right]$$

$$= N\left[1 - \frac{1}{v}l^2(k_F r_{12}) \right] - \frac{\rho}{v}\int d\vec{r}_3\, l^2(k_F r_{13}) - \frac{\rho}{v}\int d\vec{r}_3 l^2(k_F r_{23})$$

$$+ \frac{2}{v^2}l(k_F r_{12})\rho \int l(k_F r_{13})l(k_F r_{23})d\vec{r}_3$$

$$= N g_2^{FS}(r_{12}) - \frac{\rho}{v}\frac{v}{\rho} - \frac{\rho}{v}\frac{v}{\rho} + \frac{2}{v^2}v l(k_F r_{12})l(k_F r_{12}) = N g_2^{FS}(r_{12}) - 2 + 2\frac{l^2(k_F r)}{v}$$

$$= N g_2^{FS}(r_{12}) - 2\, g_2^{FS}(r_{12}) = (N-2)g_2^{FS}(r_{12}).$$

PROBLEM 11.3
Properties of the two-body density matrix.

The two-body density matrix of an N-body system is defined as

$$\rho_2(\vec{r}_1,\vec{r}_2;\vec{r}_1{}',\vec{r}_2') = N(N-1)\int d\vec{r}_3...d\vec{r}_N\, \Psi^*(\vec{r}_1,...,\vec{r}_N)\, \Psi(\vec{r}_1{}',\vec{r}_2',\vec{r}_3,...,\vec{r}_N),$$

where $\Psi(\vec{r}_1,...,\vec{r}_N)$ is a unit-normalized pure state, and where we have omitted
the spin degrees of freedom. It is also common to define the semidiagonal:

$$\rho_2^{sd}(\vec{r}_1,\vec{r}_2;\vec{r}_1{}') = \rho_2(\vec{r}_1,\vec{r}_2;\vec{r}_1{}',\vec{r}_2).$$

In the particular case of the homogeneous free Fermi sea, the two-body density
matrix is given by

$$\rho_2(\vec{r}_1,\vec{r}_2;\vec{r}_1{}',\vec{r}_2') = \rho^2\left(l(k_F r_{11'})\, l(k_F r_{22'}) - \frac{l(k_F r_{12'})\, l(k_F r_{21'})}{v} \right),$$

where $l(z)$ is the Slater function, v is the spin degeneracy and $r_{ab'} = |\vec{r}_a - \vec{r}_b'|$.
Notice that the spin degrees of freedom have been summed up.
Demonstrate the following properties:

(a)
$$\rho_2(\vec{r}_1,\vec{r}_2;\vec{r}_1,\vec{r}_2) = \rho^2 g(r_{12}).$$

(b)
$$\int \rho_2^{sd}(\vec{r}_1,\vec{r}_2;\vec{r}_1{}')\, d\vec{r}_2 = (N-1)\, \rho_1(\vec{r}_1,\vec{r}_1{}'),$$

which in the case of a homogeneous system depends only on $r_{11'}$, i.e.,

$$\rho_1(\vec{r}_1,\vec{r}_1{}') = \rho_1(r_{11'}).$$

In both cases, check the properties for the free Fermi sea.

(a) Taking into account the definition of $g(r_{12})$ for a homogeneous system,

$$g(r_{12}) = \frac{N(N-1)}{\rho^2} \int d\vec{r}_3 \dots d\vec{r}_N \, \Psi^*(\vec{r}_1, \dots \vec{r}_N) \, \Psi(\vec{r}_1, \dots \vec{r}_N),$$

it is immediate to check that

$$\rho_2(\vec{r}_1, \vec{r}_2; \vec{r}_1, \vec{r}_2) = N(N-1) \int d\vec{r}_3 \dots d\vec{r}_N \, \Psi^*(\vec{r}_1, \dots \vec{r}_N) \, \Psi(\vec{r}_1, \dots \vec{r}_N) \; = \; \rho^2 g(r_{12}).$$

For the free Fermi sea we have:

$$\rho_2(\vec{r}_1, \vec{r}_2; \vec{r}_1, \vec{r}_2) = \rho^2 \left(l(0)l(0) - \frac{1}{\nu} l(k_F r_{12}) l(k_F r_{12}) \right)$$

$$= \rho^2 \left(1 - \frac{l^2(k_F r_{12})}{\nu} \right) = \rho^2 g_{\mathrm{FS}}(r_{12}).$$

(b) For the second property,

$$\int \rho_2(\vec{r}_1, \vec{r}_2; \vec{r}_1{}', \vec{r}_2) d\vec{r}_2 = N(N-1) \int d\vec{r}_2 \int d\vec{r}_3 \dots d\vec{r}_N \, \Psi^*(\vec{r}_1, \dots \vec{r}_N) \Psi(\vec{r}_1{}', \dots \vec{r}_N)$$

$$= (N-1)\rho_1(\vec{r}_1, \vec{r}_1{}'),$$

where

$$\rho_1(\vec{r}_1, \vec{r}_1{}') = N \int d\vec{r}_2 \dots d\vec{r}_N \Psi^*(\vec{r}_1, \dots \vec{r}_N) \Psi(\vec{r}_1{}', \dots, \vec{r}_N),$$

which in the case of a homogeneous system depends only on $r_{11'} = |\vec{r}_1 - \vec{r}_1{}'|$.
In the case of the free Fermi sea,

$$\rho_{2,\mathrm{FS}}^{\mathrm{sd}}(\vec{r}_1, \vec{r}_2; \vec{r}_1{}') = \rho^2 \left(l(k_F r_{11'}) - \frac{1}{\nu} l(k_F r_{12}) \, l(k_F r_{1'2}) \right).$$

The first term does not depend on $\vec{r}_2$ and therefore can be immediately integrated,

$$\rho^2 \int l(k_F r_{11'}) \, d\vec{r}_2 = \rho^2 \, \Omega \, l(k_F r_{11'}) = N \, \rho \, l(k_F r_{11'}),$$

while for the second term we can use that

$$\rho^2 \int d\vec{r}_2 \, l(k_F r_{12}) \, l(k_F r_{1'2}) \, d\vec{r}_2 = \rho \nu l(k_F r_{11'}),$$

and therefore,

$$\int \rho_{2,\mathrm{FS}}^{\mathrm{sd}}(\vec{r}_1, \vec{r}_2; \vec{r}_1{}') \, d\vec{r}_2 = N \, \rho \, l(k_F r_{11'}) - \rho \, \nu \, \frac{l(k_F r_{11'})}{\nu}$$

$$= (N-1) \, \rho \, l(k_F r_{11'}) = (N-1) \, \rho_{1,\mathrm{FS}}(r_{11'}).$$

PROBLEM 11.4
One-body density matrix in a 2D non-interacting Bose-Hubbard gas

The non-interacting Bose-Hubbard Hamiltonian is defined as

$$\hat{H} = -J \sum_{\langle j,k \rangle} \hat{a}_j^\dagger \hat{a}_k + \hat{a}_k^\dagger \hat{a}_j,$$

where $\hat{a}_k$ ($\hat{a}_k^\dagger$) is the annihilation (creation) operator for site k and $\langle j,k \rangle$ represents that the sum is only performed over neighboring indices. For a 2D system with square lattice of spacing d and size $N_s \times N_s$, with periodic boundary conditions ($\hat{a}_{N_s+1} = \hat{a}_1$), filled with N bosons:

(a) Find the normalization coefficient for the following transformation:

$$\hat{a}_j = C(N_s) \sum_{q_x} \sum_{q_y} e^{i \vec{r}_j \cdot \vec{q}} \hat{b}_{q_x q_y},$$

where $\vec{r}_j = m_x d \, \vec{u}_x + m_y d \, \vec{u}_y$ is the position of site j ($m_x, m_y = 1, 2, \ldots, N_s$), for $\hat{b}_{q_x q_y}$ to be the annihilation operator of a particle with momentum $\vec{q} = q_x \vec{u}_x + q_y \vec{u}_y$ which fulfills the following commutation relations:

$$\left[\hat{b}_{q_x q_y}, \hat{b}_{q'_x q'_y}^\dagger \right] = \delta_{q_x q'_x} \delta_{q_y q'_y}. \tag{11.4.1}$$

(b) Diagonalize the Hamiltonian using the previous transformation.

(c) Compute the one-body density matrix elements ρ_{jk} for the ground state $|\Psi_0\rangle$ of the previous Hamiltonian,

$$\rho_{jk} = \frac{1}{N} \left\langle \Psi_0 \middle| \hat{a}_j^\dagger \hat{a}_k \middle| \Psi_0 \right\rangle.$$

(a) Let us write the commutation relations in the positions space to check what is needed for $C(N_s)$ to fulfill.

$$
\begin{aligned}
\left[\hat{a}_j, \hat{a}_k^\dagger\right] &= |C(N_s)|^2 \left(\sum_{q_x}\sum_{q_y} e^{i\vec{r}_j\cdot\vec{q}}\,\hat{b}_{q_xq_y} \sum_{q_x'}\sum_{q_y'} e^{-i\vec{r}_k\cdot\vec{q}'}\,\hat{b}_{q_x'q_y'}^\dagger \right. \\
&\left. \quad - \sum_{q_x'}\sum_{q_y'} e^{-i\vec{r}_k\cdot\vec{q}'}\,\hat{b}_{q_x'q_y'}^\dagger \sum_{q_x}\sum_{q_y} e^{i\vec{r}_j\cdot\vec{q}}\,\hat{b}_{q_xq_y} \right) \\
&= |C(N_s)|^2 \sum_{q_x}\sum_{q_y}\sum_{q_x'}\sum_{q_y'} e^{i(\vec{r}_j\cdot\vec{q}-\vec{r}_k\cdot\vec{q}')} \left(\hat{b}_{q_xq_y}\hat{b}_{q_x'q_y'}^\dagger - \hat{b}_{q_x'q_y'}^\dagger \hat{b}_{q_xq_y} \right) \\
&= |C(N_s)|^2 \sum_{q_x}\sum_{q_y}\sum_{q_x'}\sum_{q_y'} e^{i(\vec{r}_j\cdot\vec{q}-\vec{r}_k\cdot\vec{q}')} \left[\hat{b}_{q_xq_y}, \hat{b}_{q_x'q_y'}^\dagger \right].
\end{aligned}
$$

Now, using the relation in Eq. (11.4.1) we obtain:

$$
\begin{aligned}
\left[\hat{a}_j, \hat{a}_k^\dagger\right] &= |C(N_s)|^2 \sum_{q_x}\sum_{q_y}\sum_{q_x'}\sum_{q_y'} e^{i(\vec{r}_j\cdot\vec{q}-\vec{r}_k\cdot\vec{q}')} \delta_{q_xq_x'}\delta_{q_yq_y'} \\
&= |C(N_s)|^2 \sum_{q_x}\sum_{q_y} e^{i(\vec{r}_j-\vec{r}_k)\vec{q}} = |C(N_s)|^2 \sum_{q_x}\sum_{q_y} e^{i(r_{jx}-r_{kx})q_x}\, e^{i(r_{jy}-r_{ky})q_y}.
\end{aligned}
$$

Since we know that the momentum values in the first Brillouin zone are of the form

$$
\vec{q} = \frac{2\pi k_x}{dN_s}\vec{u}_x + \frac{2\pi k_y}{dN_s}\vec{u}_y, \quad k_x, k_y = -\lfloor N_s/2 \rfloor, \ldots, -1, 0, 1, \ldots, \lceil N_s/2 \rceil
$$

we are able to perform the summations using the following property of complex sums: $\sum_{k=0}^{M} e^{i2\pi k/M} = M\delta_k$. Then, ignoring a global phase that will not affect the system, we can define the normalization value:

$$
\left[\hat{a}_j, \hat{a}_k^\dagger\right] = |C(N_s)|^2 N_s^2 \delta_{jk} \implies |C(N_s)|^2 = \frac{1}{N_s^2} \implies C(N_s) = \frac{1}{N_s}.
$$

(b) First of all, let us write the Hamiltonian using the fact that the sum is only performed over first neighbors and we have a square lattice:

$$
\hat{H} = -J \sum_{m_x=1}^{N_s} \sum_{m_y=1}^{N_s} \left(\hat{a}_{m_x+1,m_y}^\dagger \hat{a}_{m_x,m_y} + \hat{a}_{m_x,m_y+1}^\dagger \hat{a}_{m_x,m_y} \right) + \text{h.c.},
$$

where h.c. stands for the Hermitian conjugate of the first terms and the creation and annihilation operators now are labeled following the spatial integer coefficients m_x, m_y that identify each lattice point.

To diagonalize the Hamiltonian, we substitute the annihilation and creation operators by their momentum series:

$$\hat{H} = -J\frac{1}{N_s^2}\sum_{m_x,m_y}\sum_{k_x,k_x',k_y,k_y'}\left(e^{i\frac{2\pi}{N_s}(m_x(k_x'-k_x)+m_y(k_y'-k_y))}e^{-i\frac{2\pi}{N_s}k_x}\hat{b}_{k_xk_y}^{\dagger}\hat{b}_{k_x'k_y'}\right.$$

$$\left.+e^{i\frac{2\pi}{N_s}(m_x(k_x'-k_x)+m_y(k_y'-k_y))}e^{-i\frac{2\pi}{N_s}k_y}\hat{b}_{k_xk_y}^{\dagger}\hat{b}_{k_x'k_y'}\right)$$

$$-J\frac{1}{N_s^2}\sum_{m_x,m_y}\sum_{k_x,k_x',k_y,k_y'}\left(e^{i\frac{2\pi}{N_s}(m_x(k_x'-k_x)+m_y(k_y'-k_y))}e^{i\frac{2\pi}{N_s}k_x'}\hat{b}_{k_xk_y}^{\dagger}\hat{b}_{k_x'k_y'}\right.$$

$$\left.+e^{i\frac{2\pi}{N_s}(m_x(k_x'-k_x)+m_y(k_y'-k_y))}e^{i\frac{2\pi}{N_s}k_y'}\hat{b}_{k_xk_y}^{\dagger}\hat{b}_{k_x'k_y'}\right)$$

$$=-\frac{J}{N_s^2}\sum_{\substack{m_x,m_y\\k_x,k_x',k_y,k_y'}}e^{i\frac{2\pi}{N_s}(m_x(k_x'-k_x)+m_y(k_y'-k_y))}\left(e^{-i\frac{2\pi}{N_s}k_x}+e^{-i\frac{2\pi}{N_s}k_y}\right.$$

$$\left.+e^{i\frac{2\pi}{N_s}k_x'}+e^{i\frac{2\pi}{N_s}k_y'}\right)\hat{b}_{k_xk_y}^{\dagger}\hat{b}_{k_x'k_y'}.$$

Now, performing the summation over the spatial indices m_x and m_y,

$$\hat{H}=-\frac{J}{N_s^2}\sum_{k_x,k_x',k_y,k_y'}N_s^2\delta_{k_xk_x'}\delta_{k_yk_y'}\left(e^{-i\frac{2\pi}{N_s}k_x}+e^{-i\frac{2\pi}{N_s}k_y}+e^{i\frac{2\pi}{N_s}k_x'}+e^{i\frac{2\pi}{N_s}k_y'}\right)\hat{b}_{k_xk_y}^{\dagger}\hat{b}_{k_x'k_y'}$$

$$=-J\sum_{k_x,k_y}2\left(\cos\left(\frac{2\pi k_x}{N_s}\right)+\cos\left(\frac{2\pi k_y}{N_s}\right)\right)\hat{b}_{k_xk_y}^{\dagger}\hat{b}_{k_xk_y}$$

$$=-J\sum_{k_x,k_y}2\left(\cos\left(\frac{2\pi k_x}{N_s}\right)+\cos\left(\frac{2\pi k_y}{N_s}\right)\right)\hat{m}_{k_xk_y},$$

where $\hat{m}_{k_xk_y}$ is the number operator in momentum space. Then this is an already diagonal Hamiltonian, whose ground state for a system with N bosons is the state with all bosons in the state $k_x=0$, $k_y=0$:

$$|\Psi_0\rangle = \frac{1}{\sqrt{N!}}\left(\hat{b}_{00}^{\dagger}\right)^N|\text{vac}\rangle.$$

(c) First, note the following relations:

$$\underbrace{\hat{a}^{\dagger}\hat{a}^{\dagger}\ldots\hat{a}^{\dagger}}_{l\text{ times}}\underbrace{\hat{a}\hat{a}\ldots\hat{a}}_{l\text{ times}}=\hat{n}\left(\hat{n}-1\right)\left(\hat{n}-2\right)\cdots\left(\hat{n}-(l-1)\right),\tag{11.4.2}$$

$$\underbrace{\hat{a}\hat{a}\ldots\hat{a}}_{l\text{ times}}\underbrace{\hat{a}^{\dagger}\hat{a}^{\dagger}\ldots\hat{a}^{\dagger}}_{l\text{ times}}=\left(\hat{n}+1\right)\left(\hat{n}+2\right)\cdots\left(\hat{n}+l\right).\tag{11.4.3}$$

Now we compute the one-body density matrix elements ρ_{jk}:

$$
\begin{aligned}
\rho_{jk} &= \frac{1}{NN!} \langle \text{vac}| \left(\hat{b}_{00}\right)^N a_j^\dagger a_k \left(\hat{b}_{00}^\dagger\right)^N |\text{vac}\rangle \\
&= \frac{1}{NN!N_s^2} \langle \text{vac}| \left(\hat{b}_{00}\right)^N \sum_{q_x}\sum_{q_y} e^{-i\vec{r}_j\cdot\vec{q}} \hat{b}_{q_x q_y}^\dagger \sum_{q_x'}\sum_{q_y'} e^{i\vec{r}_k\cdot\vec{q}'} \hat{b}_{q_x' q_y'} \left(\hat{b}_{00}^\dagger\right)^N |\text{vac}\rangle \\
&= \frac{1}{NN!N_s^2} \langle \text{vac}| \left(\hat{b}_{00}\right)^N \left(\hat{b}_{00}^\dagger + \sum_{\substack{q_x,q_y \\ q_x,q_y\neq 0,0}} e^{-i\vec{r}_j\cdot\vec{q}} \hat{b}_{q_x q_y}^\dagger \right) \\
&\quad \times \left(\hat{b}_{00} + \sum_{\substack{q_x',q_y' \\ q_x',q_y'\neq 0,0}} e^{i\vec{r}_k\cdot\vec{q}'} \hat{b}_{q_x' q_y'} \right) \left(\hat{b}_{00}^\dagger\right)^N |\text{vac}\rangle .
\end{aligned}
$$

Taking into account that the terms under the sum commute with the operators that act on the vacuum state, it is clear that they do not contribute to this computation. Then, the expression reduces to

$$
\begin{aligned}
\rho_{jk} &= \frac{1}{NN!N_s^2} \langle \text{vac}| \left(\hat{b}_{00}\right)^N \hat{b}_{00}^\dagger \hat{b}_{00} \left(\hat{b}_{00}^\dagger\right)^N |\text{vac}\rangle \\
&= \frac{1}{NN!N_s^2} \langle \text{vac}| \left(\hat{b}_{00}\right)^{N+1} \left(\hat{b}_{00}^\dagger\right)^{N+1} - \left(\hat{b}_{00}\right)^N \left(\hat{b}_{00}^\dagger\right)^N |\text{vac}\rangle ,
\end{aligned}
$$

and applying relation Eq. (11.4.3) we have:

$$
\begin{aligned}
\rho_{jk} &= \frac{1}{NN!N_s^2} \langle \text{vac}| (\hat{m}_{00}+1)\cdots(\hat{m}_{00}+N+1) - (\hat{m}_{00}+1)\cdots(\hat{m}_{00}+N) |\text{vac}\rangle \\
&= \frac{(N+1)!-N!}{NN!N_s^2} = \frac{N!(N+1-1)}{NN!N_s^2} = \frac{1}{N_s^2} .
\end{aligned}
$$

The general expression is found to be independent of the indices j,k. This is a feature that highlights the non-decaying correlations of the ground state between different sites.

12 Superfluidity

Superfluidity is one of the most striking manifestations of quantum mechanics at macroscopic scales, characterized by frictionless flow and the emergence of collective quantum phenomena. The theoretical foundation of superfluidity in fermionic systems was revolutionized by the Bardeen-Cooper-Schrieffer (BCS) theory , developed in 1957 to explain superconductivity in metals. This groundbreaking work provided a microscopic description of how attractive interactions between electrons near the Fermi surface can lead to the formation of Cooper pairs—correlated pairs of fermions that condense into a coherent quantum state. The BCS theory not only explained superconductivity but also laid the groundwork for understanding superfluidity in systems like liquid helium-3 and ultracold fermionic gases, becoming a cornerstone of condensed matter physics.

This chapter explores the BCS state through a series of problems designed to introduce the reader to the BCS theory. The BCS wavefunction captures the essence of superfluidity by describing a macroscopic quantum state with off-diagonal long-range order. Readers will engage with the mathematical structure of the BCS state and calculate some properties. Through this exploration, the chapter underscores the profound impact of the BCS framework on both theoretical physics and technological advancements.

PROBLEM 12.1
The Bardeen-Cooper-Schrieffer (BCS) ansatz

The celebrated BCS ansatz reads

$$|\text{BCS}\rangle = \prod_{k>0} \left(u_k + v_k \hat{a}_k^\dagger \hat{a}_{-k}^\dagger \right) |\text{vac}\rangle, \quad u_k, v_k \in \mathbb{C},$$

where k denotes all quantum numbers necessary to describe a state, and the minus sign in front means time-reversed. The notation $k > 0$ in the product refers to the positive eigenvalue of time reversal.

(a) Derive the normalization equations for u_k and v_k, and show that $|u_k|^2 + |v_k|^2 = 1 \, \forall k$ is a valid choice. Use this condition for the next sections.

(b) Compute the average particle number of a BCS state in terms of the constants u_k, v_k.

(c) Compute the fluctuations in the particle number operator, $\left(\Delta \hat{N} \right)^2$.

(a) We can start by imposing that the BCS state have unit norm,

$$1 = \langle \text{BCS} | \text{BCS} \rangle.$$

Then, using the definition of the BCS state:

$$1 = \langle \text{vac} | \prod_{k'>0} \left(\bar{u}_{k'} + \bar{v}_{k'} \hat{a}_{-k'} \hat{a}_{k'} \right) \prod_{k>0} \left(u_k + v_k \hat{a}_k^\dagger \hat{a}_{-k}^\dagger \right) |\text{vac}\rangle.$$

We can now bring together the pairs of factors labeled the same, i.e., $k' = k$, and write the equation above as a single product over k. To see this, notice that factors within each product commute. Furthermore, the k-th factor in the right product also commutes with all factors in the left product except for the one with $k' = k$, and therefore we can bring these two factors together, respecting their initial order:

$$1 = \langle \text{vac} | \prod_{k>0} \left(\bar{u}_k + \bar{v}_k \hat{a}_{-k} \hat{a}_k \right) \left(u_k + v_k \hat{a}_k^\dagger \hat{a}_{-k}^\dagger \right) |\text{vac}\rangle$$

$$= \langle \text{vac} | \prod_{k>0} \left(|u_k|^2 + \underbrace{\bar{u}_k v_k \hat{a}_k^\dagger \hat{a}_{-k}^\dagger}_{\text{(I)}} + \underbrace{\bar{v}_k u_k \hat{a}_{-k} \hat{a}_k}_{\text{(II)}} + |v_k|^2 \underbrace{\hat{a}_{-k} \hat{a}_k \hat{a}_k^\dagger \hat{a}_{-k}^\dagger}_{\text{(III)}} \right) |\text{vac}\rangle.$$

Here, the term (I) is zero because $\langle 0 | \hat{a}_k^\dagger = 0 \, \forall k$. The term (II) is also null, since $\hat{a}_k |0\rangle = 0 \, \forall k$. The braced factor (III) is the identity. Then,

$$1 = \prod_{k>0} \left(|u_k|^2 + |v_k|^2 \right) .$$

One way to fulfill this condition is to have:

$$|u_k|^2 + |v_k|^2 = 1 \quad \forall k .$$

(b) We start by inserting the definition of the particle number operator $\hat{N}$ in the expectation value:

$$
\begin{aligned}
\langle \hat{N} \rangle_{\mathrm{BCS}} &= \langle \mathrm{BCS} | \hat{N} | \mathrm{BCS} \rangle \\
&= \langle \mathrm{vac} | \prod_{k'>0} (\bar{u}_{k'} + \bar{v}_{k'} \hat{a}_{-k'} \hat{a}_{k'}) \sum_q \hat{a}_q^\dagger \hat{a}_q \prod_{k>0} \left(u_k + v_k \hat{a}_k^\dagger \hat{a}_{-k}^\dagger \right) | \mathrm{vac} \rangle \\
&= \sum_q \langle \mathrm{vac} | \prod_{k'>0} (\bar{u}_{k'} + \bar{v}_{k'} \hat{a}_{k'} \hat{a}_{-k'}) \hat{a}_q^\dagger \hat{a}_q \prod_{k>0} \left(u_k + v_k \hat{a}_k^\dagger \hat{a}_{-k}^\dagger \right) | \mathrm{vac} \rangle .
\end{aligned}
$$

Now we cannot bring equally-labeled terms together, since the factors $\hat{a}_q^\dagger \hat{a}_q$ will not commute with the factors in the products when $k, k' = q$. What we can then do is the following:

$$
\begin{aligned}
\langle \hat{N} \rangle_{\mathrm{BCS}} &= \sum_q \langle \mathrm{vac} | \prod_{k' \neq q} (\bar{u}_{k'} + \bar{v}_{k'} \hat{a}_{-k'} \hat{a}_{k'}) (\bar{u}_q + \bar{v}_q \hat{a}_{-q} \hat{a}_q) \hat{a}_q^\dagger \hat{a}_q \left(u_q + v_q \hat{a}_q^\dagger \hat{a}_{-q}^\dagger \right) \\
&\quad \times \prod_{k \neq q} \left(u_k + v_k \hat{a}_k^\dagger \hat{a}_{-k}^\dagger \right) | \mathrm{vac} \rangle \\
&= \sum_q \Bigg[\langle \mathrm{vac} | \prod_{k' \neq q} |u_q|^2 (\bar{u}_{k'} + \bar{v}_{k'} \hat{a}_{-k'} \hat{a}_{k'}) \hat{a}_q^\dagger \hat{a}_q \prod_{k \neq q} \left(u_k + v_k \hat{a}_k^\dagger \hat{a}_{-k}^\dagger \right) | \mathrm{vac} \rangle \\
&\quad + \langle \mathrm{vac} | \prod_{k' \neq q} \bar{u}_q v_q (\bar{u}_{k'} + \bar{v}_{k'} \hat{a}_{-k'} \hat{a}_{k'}) \hat{a}_q^\dagger \hat{a}_q \hat{a}_q^\dagger \hat{a}_{-q}^\dagger \prod_{k \neq q} \left(u_k + v_k \hat{a}_k^\dagger \hat{a}_{-k}^\dagger \right) | \mathrm{vac} \rangle \\
&\quad + \langle \mathrm{vac} | \prod_{k' \neq q} \bar{v}_q u_q (\bar{u}_{k'} + \bar{v}_{k'} \hat{a}_{-k'} \hat{a}_{k'}) \hat{a}_{-q} \hat{a}_q \hat{a}_q^\dagger \hat{a}_q \prod_{k \neq q} \left(u_k + v_k \hat{a}_k^\dagger \hat{a}_{-k}^\dagger \right) | \mathrm{vac} \rangle \\
&\quad + \langle \mathrm{vac} | \prod_{k' \neq q} |v_q|^2 (\bar{u}_{k'} + \bar{v}_{k'} \hat{a}_{-k'} \hat{a}_{k'}) \hat{a}_{-q} \hat{a}_q \hat{a}_q^\dagger \hat{a}_q \hat{a}_q^\dagger \hat{a}_{-q}^\dagger \\
&\quad \times \prod_{k \neq q} \left(u_k + v_k \hat{a}_k^\dagger \hat{a}_{-k}^\dagger \right) | \mathrm{vac} \rangle \Bigg] .
\end{aligned}
$$

The first term is immediately zero, since the operators $\hat{a}_q$ commute with the whole product to their right, and they will then annihilate the kets $|0\rangle$. With similar arguments, one can see that the same happens for the next two terms, leaving us only with the fourth term, which we can rewrite as a single product. Ignoring the crossed terms (those with factors $\bar{u}_k v_k$ and $\bar{v}_k u_k$) because they also vanish following the arguments of the previous question, the result is:

$$\langle \hat{N} \rangle_{\text{BCS}} = \sum_q |v_q|^2 \prod_{k \neq q} \Big(|u_k|^2 \, \langle \text{vac} \, | \hat{a}_{-q} \hat{a}_q \hat{a}_q^\dagger \hat{a}_q \hat{a}_q^\dagger \hat{a}_{-q}^\dagger | \, \text{vac} \rangle$$

$$+ |v_k|^2 \, \langle \text{vac} \, | \hat{a}_{-k} \hat{a}_k \hat{a}_{-q} \hat{a}_q \hat{a}_q^\dagger \hat{a}_q \hat{a}_q^\dagger \hat{a}_{-q}^\dagger \hat{a}_k^\dagger \hat{a}_{-k}^\dagger | \, \text{vac} \rangle \Big)$$

$$= \sum_q |v_q|^2 \prod_{k \neq q} \Big(|u_k|^2 \, \langle \text{vac} \, | \hat{a}_{-q} \hat{a}_q \hat{a}_q^\dagger \hat{a}_q \hat{a}_q^\dagger \hat{a}_{-q}^\dagger | \, \text{vac} \rangle$$

$$+ |v_k|^2 \, \langle \text{vac} \, | \hat{a}_{-k} \hat{a}_k \hat{a}_{-q} \hat{a}_q \hat{a}_q^\dagger \hat{a}_q \hat{a}_q^\dagger \hat{a}_{-q}^\dagger \hat{a}_k^\dagger \hat{a}_{-k}^\dagger | \, \text{vac} \rangle \Big)$$

$$= \sum_q |v_q|^2 \prod_{k \neq q} \big(|u_k|^2 + |v_k|^2 \big) = \sum_q \frac{|v_q|^2}{|u_q|^2 + |v_q|^2} = \sum_q |v_q|^2,$$

where we have used Wick's theorem, and the normalization condition $|u_k|^2 + |v_k|^2 = 1 \, \forall k$ has been used in the last step. It is interesting to note that the average number of particles in a BCS ansatz depends only on the $\{v_k\}_k$.[1]

(c) The fluctuation in the number of particles, $\Delta \hat{N}$, is defined as the variance of $\hat{N}$, $\Delta \hat{N} = \langle \hat{N}^2 \rangle - \langle \hat{N} \rangle^2$. The second term is trivially

$$\langle \hat{N} \rangle^2 = \sum_{q,q'} |v_q|^2 |v_{q'}|^2,$$

whereas the first term requires some more work. By definition,

$$\hat{N}^2 = \sum_{q,q'} \hat{a}_{q'}^\dagger \hat{a}_{q'} \hat{a}_q^\dagger \hat{a}_q,$$

and therefore we can write:

$$\langle \hat{N}^2 \rangle_{\text{BCS}} = \sum_{q,q'} \langle \text{vac} \, | \prod_{k'>0} (\bar{u}_{k'} + \bar{v}_{k'} \hat{a}_{-k'} \hat{a}_{k'}) \, \hat{a}_{q'}^\dagger \hat{a}_{q'} \hat{a}_q^\dagger \hat{a}_q \prod_{k>0} \Big(u_k + v_k \hat{a}_k^\dagger \hat{a}_{-k}^\dagger \Big) | \, \text{vac} \rangle \, .$$

Here, in order to be able to use the trick of pulling out the factors with $k, k' = q, q'$, we must first differentiate the cases when $q' = q$ and $q' \neq q$:

$$\sum_{q,q'} = \sum_q \Big([q' = q] + \sum_{q' \neq q} \Big).$$

---

[1] In the literature, this result is often expressed as a sum over $q > 0$, and with an extra factor of 2 to exploit the time-reversal symmetry of the BCS ansatz.

Let us start with the terms where $q' = q$:

$$\sum_q \langle \text{vac} | \prod_{k'>0} (\bar{u}_{k'} + \bar{v}_{k'} \hat{a}_{-k'} \hat{a}_{k'}) \hat{a}_q^\dagger \hat{a}_q \hat{a}_q^\dagger \hat{a}_q \prod_{k>0} \left(u_k + v_k \hat{a}_k^\dagger a_{-k}^\dagger \right) | \text{vac} \rangle$$

$$= \sum_q \langle \text{vac} | \prod_{k' \neq q} (\bar{u}_{k'} + \bar{v}_{k'} \hat{a}_{-k'} \hat{a}_{k'}) (\bar{u}_q + \bar{v}_q \hat{a}_{-q} \hat{a}_q) \hat{a}_q^\dagger \hat{a}_q \hat{a}_q^\dagger \hat{a}_q \left(u_q + v_q \hat{a}_q^\dagger a_{-q}^\dagger \right)$$

$$\times \prod_{k \neq q} \left(u_k + v_k \hat{a}_k^\dagger a_{-k}^\dagger \right) | \text{vac} \rangle$$

$$= \sum_q \langle \text{vac} | \prod_{k' \neq q} |v_q|^2 (\bar{u}_{k'} + \bar{v}_{k'} \hat{a}_{-k'} \hat{a}_{k'}) \hat{a}_{-q} \hat{a}_q \hat{a}_q^\dagger \hat{a}_q \hat{a}_q^\dagger \hat{a}_q \hat{a}_q^\dagger \hat{a}_{-q}^\dagger$$

$$\times \prod_{k \neq q} \left(u_k + v_k \hat{a}_k^\dagger a_{-k}^\dagger \right) | \text{vac} \rangle$$

$$= \sum_q |v_q|^2 \prod_{k \neq q} \langle \text{vac} | (\bar{u}_k + \bar{v}_k \hat{a}_{-k} \hat{a}_k) \hat{a}_{-q} \hat{a}_q \hat{a}_q^\dagger \hat{a}_q \hat{a}_q^\dagger \hat{a}_q \hat{a}_q^\dagger \hat{a}_{-q}^\dagger \left(u_k + v_k \hat{a}_k^\dagger a_{-k}^\dagger \right) | \text{vac} \rangle$$

$$= \sum_q |v_q|^2 \prod_{k \neq q} \left(|u_k|^2 \langle \text{vac} | \hat{a}_{-q} \hat{a}_q \hat{a}_q^\dagger \hat{a}_q \hat{a}_q^\dagger \hat{a}_q \hat{a}_q^\dagger \hat{a}_{-q}^\dagger | \text{vac} \rangle \right.$$

$$\left. + |v_k|^2 \langle \text{vac} | \hat{a}_{-k} \hat{a}_k \hat{a}_{-q} \hat{a}_q \hat{a}_q^\dagger \hat{a}_q \hat{a}_q^\dagger \hat{a}_q \hat{a}_q^\dagger \hat{a}_{-q}^\dagger \hat{a}_k^\dagger \hat{a}_{-k}^\dagger | \text{vac} \rangle \right)$$

$$= \sum_q |v_q|^2 \prod_{k \neq q} \left(|u_k|^2 + |v_k|^2 \right) = \sum_q \frac{|v_q|^2}{|u_q|^2 + |v_q|^2} \, .$$

Let us now compute the terms where $q' \neq q$:

$$\sum_q \sum_{q' \neq q} \langle \text{vac} | \prod_{k'>0} (\bar{u}_{k'} + \bar{v}_{k'} \hat{a}_{-k'} \hat{a}_{k'}) \hat{a}_{q'}^\dagger \hat{a}_{q'} \hat{a}_q^\dagger \hat{a}_q \prod_{k>0} \left(u_k + v_k \hat{a}_k^\dagger \hat{a}_{-k}^\dagger \right) | \text{vac} \rangle$$

$$= \sum_q \sum_{q' \neq q} \langle \text{vac} | \prod_{k' \neq q} (\bar{u}_{k'} + \bar{v}_{k'} \hat{a}_{-k'} \hat{a}_{k'}) (\bar{u}_q + \bar{v}_q \hat{a}_{-q} \hat{a}_q) \hat{a}_{q'}^\dagger \hat{a}_{q'} \hat{a}_q^\dagger \hat{a}_q$$

$$\times \left(u_q + v_q \hat{a}_q^\dagger \hat{a}_{-q}^\dagger \right) \prod_{k \neq q} \left(u_k + v_k \hat{a}_k^\dagger \hat{a}_{-k}^\dagger \right) | \text{vac} \rangle$$

$$= \sum_q \sum_{q' \neq q} \langle \text{vac} | \prod_{k' \neq q} |v_q|^2 (\bar{u}_{k'} + \bar{v}_{k'} \hat{a}_{-k'} \hat{a}_{k'}) \hat{a}_{-q} \hat{a}_q \hat{a}_{q'}^\dagger \hat{a}_{q'} \hat{a}_q^\dagger \hat{a}_q \hat{a}_q^\dagger \hat{a}_{-q}^\dagger$$

$$\times \prod_{k \neq q} \left(u_k + v_k \hat{a}_k^\dagger \hat{a}_{-k}^\dagger \right) | \text{vac} \rangle \, .$$

Here, notice that for every q we have $q' \neq q$, and also the products do not contain factors labelled by q. This means that the factors $\hat{a}_q$, $\hat{a}_{-q}$ and their adjoints must be contracted with themselves, and there is only one such allowed Wick contraction. This leaves us with:

$$
\sum_q \sum_{q' \neq q} |v_q|^2 \langle \text{vac} | \prod_{k' \neq q} (\bar{u}_{k'} + \bar{v}_{k'} \hat{a}_{-k'} \hat{a}_{k'}) \hat{a}_{q'}^\dagger \hat{a}_{q'} \prod_{k \neq q} \left(u_k + v_k \hat{a}_k^\dagger \hat{a}_{-k}^\dagger \right) | \text{vac} \rangle
$$

$$
= \sum_q \sum_{q' \neq q} |v_q|^2 \langle \text{vac} | \prod_{k' \neq q, q'} (\bar{u}_{k'} + \bar{v}_{k'} \hat{a}_{-k'} \hat{a}_{k'}) \left(\bar{u}_{q'} + \bar{v}_{q'} \hat{a}_{-q'} \hat{a}_{q'} \right) \hat{a}_{q'}^\dagger \hat{a}_{q'}
$$

$$
\times \left(u_{q'} + v_{q'} \hat{a}_{q'}^\dagger \hat{a}_{-q'}^\dagger \right) \prod_{k \neq q, q'} \left(u_k + v_k \hat{a}_k^\dagger \hat{a}_{-k}^\dagger \right) | \text{vac} \rangle
$$

$$
= \sum_q \sum_{q' \neq q} |v_q|^2 |v_{q'}|^2 \langle \text{vac} | \prod_{k \neq q, q'} (\bar{u}_k + \bar{v}_k \hat{a}_{-k} \hat{a}_k) \hat{a}_{-q'} \hat{a}_{q'} \hat{a}_{q'}^\dagger \hat{a}_{q'} \hat{a}_{q'}^\dagger \hat{a}_{-q'}^\dagger
$$

$$
\times \left(u_k + v_k \hat{a}_k^\dagger \hat{a}_{-k}^\dagger \right) | \text{vac} \rangle
$$

$$
= \sum_q \sum_{q' \neq q} |v_q|^2 |v_{q'}|^2 \prod_{k \neq q, q'} \left(|u_k|^2 + |v_k|^2 \right) = \sum_q \sum_{q' \neq q} \frac{|v_q|^2}{|u_q|^2 + |v_q|^2} \frac{|v_{q'}|^2}{|u_{q'}|^2 + |v_{q'}|^2},
$$

where we have used contraction lines to indicate the only possible contraction of the operators $\hat{a}_{q'}$, $\hat{a}_{-q'}$ and their adjoints. Bringing both contributions together ($q' = q$ and $q' \neq q$) yields:

$$
\langle \hat{N}^2 \rangle_{\text{BCS}} = \sum_q \left(\frac{|v_q|^2}{|u_q|^2 + |v_q|^2} + \sum_{q' \neq q} \frac{|v_q|^2}{|u_q|^2 + |v_q|^2} \frac{|v_{q'}|^2}{|u_{q'}|^2 + |v_{q'}|^2} \right).
$$

Finally,

$$
\Delta \hat{N} = \langle \hat{N}^2 \rangle - \langle \hat{N} \rangle^2 = \sum_q \left(\frac{|v_q|^2}{|u_q|^2 + |v_q|^2} + \sum_{q' \neq q} \frac{|v_q|^2}{|u_q|^2 + |v_q|^2} \frac{|v_{q'}|^2}{|u_{q'}|^2 + |v_{q'}|^2} \right.
$$

$$
\left. - \sum_{q'} \frac{|v_q|^2}{|u_q|^2 + |v_q|^2} \frac{|v_{q'}|^2}{|u_{q'}|^2 + |v_{q'}|^2} \right)
$$

$$
= \sum_q \left(\frac{|v_q|^2}{|u_q|^2 + |v_q|^2} + \sum_{q' \neq q} \frac{|v_q|^2}{|u_q|^2 + |v_q|^2} \frac{|v_{q'}|^2}{|u_{q'}|^2 + |v_{q'}|^2} \right.
$$

$$
\left. - \sum_{q' \neq q} \frac{|v_q|^2}{|u_q|^2 + |v_q|^2} \frac{|v_{q'}|^2}{|u_{q'}|^2 + |v_{q'}|^2} - \frac{|v_q|^4}{(|u_q|^2 + |v_q|^2)^2} \right)
$$

$$
= \sum_q \frac{|v_q|^2}{|u_q|^2 + |v_q|^2} \left(1 - \frac{|v_q|^2}{|u_q|^2 + |v_q|^2} \right) = \sum_q |v_q|^2 |u_q|^2,
$$

where we have used $|u_k|^2 + |v_k|^2 = 1 \, \forall k$ in the last step.

Notice that we could have used the time-reversal symmetry of the BCS ansatz in our favor: since particles are created in pairs with quantum numbers $(k, -k)$, the number of particles in a state k will be the same as that in the state $-k$. Therefore, we could run the sum exclusively over states of positive time-reversal parity and account for the missing factor of 2,

$$\hat{N} = 2 \sum_{q>0} \hat{a}_q^\dagger \hat{a}_q \,,$$

and also

$$\hat{N}^2 = 4 \sum_{q,q'>0} \hat{a}_{q'}^\dagger \hat{a}_{q'} \hat{a}_q^\dagger \hat{a}_q \,,$$

which are the usual expressions found in the literature.

PROBLEM 12.2
Particle number projection

In a previous problem we have seen the BCS ansatz which is manifestly not an eigenstate of the particle number operator, $\hat{N}$:

$$|\text{BCS}\rangle = \prod_{k>0} \left(u_k + v_k \hat{a}_k^\dagger \hat{a}_{-k}^\dagger \right) |\text{vac}\rangle \,, \quad u_k, v_k \in \mathbb{C},$$

where k denotes all quantum numbers necessary to describe a state, and the minus sign in front means time-reversed. In practice, however, these states are still used as ansätze for many-body wavefunctions of definite particle number. This is possible because we can project onto eigenstates of $\hat{N}$.

(a) Rewrite the BCS wavefunction as a sum over the eigenstates of $\hat{N}$.

(b) The particle number projection operator reads

$$\hat{P}^{N_0} = \frac{1}{2\pi} \int_0^{2\pi} d\varphi \, e^{i\varphi(\hat{N}-N_0)} \,,$$

where N_0 is the desired eigenstate of $\hat{N}$. Prove that this is indeed a projector and that it actually works as intended, that is, that it projects any state into its N_0 component.

(a) First, we have to express the product of sums appearing in the BCS ansatz as a sum of products. To do that, we can consider the general product of binary sums

$$\prod_{k=1}^N (a_k + b_k) \,.$$

In the cases $N = 2$ and $N = 3$ we can explicitly write:

$$\prod_{k=1}^{2}(a_k + b_k) = a_1 a_2 \qquad\qquad \text{(2 a's, 0 b's)}$$

$$+ a_1 b_2 + a_2 b_1 \qquad\qquad \text{(1 a, 1 b)}$$
$$+ b_1 b_2 \qquad\qquad \text{(0 a's, 2 b's)}$$

$$\prod_{k=1}^{3}(a_k + b_k) = a_1 a_2 a_3 \qquad\qquad \text{(3 a's, 0 b's)}$$

$$+ a_1 a_2 b_3 + a_1 a_3 b_2 + a_2 a_3 b_1 \qquad\qquad \text{(2 a's, 1 b)}$$
$$+ a_3 b_1 b_2 + a_1 b_2 b_3 + a_2 b_1 b_3 \qquad\qquad \text{(1 a, 2 b's)}$$
$$+ b_1 b_2 b_3 \qquad\qquad \text{(0 a's, 3 b's).}$$

Every term has N factors, and the number of a's is complementary to the number of b's, each ranging from 0 to N. Furthermore, the subindices of the factors cannot repeat within the same term. We can express this fact as follows. Consider the set $\Omega_M \equiv \{0, 1, 2, \ldots, M\}$. Then:

$$\prod_{k=1}^{M}(a_k + b_k) = \sum_{S \subseteq \Omega_M}\left(\prod_{k \in S^c} a_k \prod_{k \in S} b_k\right), \qquad (12.2.1)$$

where the complement of S is taken with respect to Ω_M. In the case of the BCS ansatz, M denotes the number of single-particle states available in the system. Using Eq. (12.2.1) we can now rewrite the BCS ansatz for a system with M available states as:

$$|BCS\rangle = \sum_{S \subseteq \Omega_M}\left(\prod_{k \in S^c} u_k \prod_{k \in S} v_k \hat{a}_k^\dagger \hat{a}_{-k}^\dagger\right)|vac\rangle$$

$$= \sum_{S \subseteq \Omega_M}\prod_{k \in S^c} u_k \prod_{k \in S} v_k |k - k\rangle$$

$$= \sum_{S \subseteq \Omega_M}\prod_{k \in S^c} u_k \prod_{k \in S} v_k |N = 2|S|\rangle,$$

where $|S|$ is the cardinal number of the subset S. What we want is to have a sum over N. To achieve this, we can write the sum over subsets as a sum over the possible cardinal numbers of these subsets, and for each cardinal number, a sum over the subsets with this cardinality:

$$|BCS\rangle = \sum_{C=0,1,\ldots,M}\sum_{S \subseteq \Omega_M \text{ s.t. } |S|=C}\prod_{k \in S^c} u_k \prod_{k \in S} v_k |N = 2C\rangle.$$

Finally, we can rewrite the above as:

$$|BCS\rangle = \sum_{N=0,2,\ldots,2M}\left(\sum_{S \subseteq \Omega_M \text{ s.t. } 2|S|=C}\prod_{k \in S^c} u_k \prod_{k \in S} v_k\right)|N\rangle$$

$$\equiv \sum_{N=0,2,\ldots,2M} a_N |N\rangle,$$

or, more correctly,

$$|\text{BCS}\rangle = \bigoplus_{N=0,2,\dots,2M} a_N |N\rangle \,,$$

since, by construction of the Fock space, states of different particle numbers cannot be entangled. Similarly, the product symbols used above should actually be tensor products, since they relate states in the same Hilbert space (i.e., with an equal number of particles).

(b) A projector $\hat{P}$ in quantum mechanics has to be (I) idempotent ($\hat{P}^2 = \hat{P}$), and (II) self-adjoint ($\hat{P}^\dagger = \hat{P}$). Let us start by checking the second property.

$$\left(\hat{P}^{N_0}\right)^\dagger = \frac{1}{2\pi} \int_0^{2\pi} d\varphi\, e^{-i\varphi\left(\hat{N}^\dagger - N_0\right)} = \frac{1}{2\pi} \int_0^{2\pi} d\varphi\, e^{-i\varphi\left(\hat{N} - N_0\right)}$$
$$= \frac{1}{2\pi} \int_0^{2\pi} d\varphi\, e^{i\varphi\left(\hat{N} - N_0\right)} = \hat{P}^{N_0} \,,$$

where we have used the fact that $\hat{N}^\dagger = \hat{N}$, and also the fact that, in the basis where $\hat{N}$ is diagonal, $\hat{P}^{N_0} = \delta_{N,N_0} = \delta_{N_0,N}$. Using this identification of $\hat{P}^{N_0}$ with the Kronecker delta, proving the first property, $\hat{P}^2 = \hat{P}$, is immediate.

We should now check that it actually projects a state onto a subspace of states with a definite number of particles. To do this, consider a general state which is a superposition of states with different particle numbers,

$$|\psi\rangle = \sum_{N=0}^{\infty} a_N |N\rangle \,.$$

Then, applying $\hat{P}^{N_0}$ onto this state,

$$\hat{P}^{N_0} |\psi\rangle = \sum_{N=0}^{\infty} a_N \frac{1}{2\pi} \int_0^{2\pi} d\varphi\, e^{i\varphi\left(\hat{N} - N_0\right)} |N\rangle \,.$$

Since $\hat{N}$ is diagonal in the basis of particle number eigenstates, $\hat{N}|N\rangle = N|N\rangle$, we can immediately identify a Kronecker delta in the equation above:

$$\hat{P}^{N_0} |\psi\rangle = \sum_{N=0}^{\infty} a_N \underbrace{\frac{1}{2\pi} \int_0^{2\pi} d\varphi\, e^{i\varphi(N - N_0)}}_{\delta_{N,N_0}} |N\rangle = a_{N_0} |N_0\rangle \,.$$

These projectors are especially useful in many-body calculations where the wavefunction is written as a superposition of definite particle number states, such as the BCS wavefunctions (first part of this exercise). This is because, for most physical systems, we want to work with a definite number of particles, and therefore we need to project onto subspaces with the concrete number of particles that we want.

13 Mixed statistics

In Chapter 2 we saw that the Fock space is defined as:

$$\mathscr{F} = \bigoplus_{N=0}^{\infty} \mathscr{H}(N),$$

where $\mathscr{H}(N)$ are Hilbert spaces with N particles. This construction is valid when dealing with a single species of particles. There are times, however, when we have to deal with systems of different types of particles, such as fermions and bosons. In this case, the construction of the Fock space of bosons and fermions is:

$$\mathscr{F} = \left[\bigoplus_{N=0}^{\infty} \mathscr{H}_b(N)\right] \otimes \left[\bigoplus_{N=0}^{\infty} \mathscr{H}_f(N)\right],$$

where the subscripts b and f are for bosonic and fermionic Hilbert spaces, respectively. This tells us that we can have mixed (entangled) states containing both bosons and fermions, for instance. The same construction can be applied for systems with more than two distinct particle species. In this chapter we provide a working example of a system containing a mixture of one fermion and two bosons.

PROBLEM 13.1
Mixed statistics: a system of fermions and bosons

Consider a system consisting of one fully-polarized fermion and two spin-0 bosons interacting via the following Hamiltonian:

$$\hat{H} = \hat{T} + \hat{V}_0 + \hat{V}_{\text{BF}},$$

where $\hat{T}$ and $\hat{V}_0$ contain only one-body operators, and $\hat{V}_{\text{BF}}$ is a two-body operator which describes the interaction between the fermion and one boson.

(a) Write the general expression of $\hat{H}$ in terms of fermionic and bosonic operators.

(b) Consider $\hat{V}_{\text{BF}} = 0$ for now. We will suppose that there are only two single-particle states available, both for fermions and for bosons, with associated single-particle energies $e_{0,f,b}, e_{1,f,b}$ fulfilling $e_{0,f,b} < e_{1,f,b}$ and $e_{1,f} < e_{1,b}$. Write the ground state of $\hat{H}$ in this case and compute its energy, E_0.

(c) Consider now $\hat{V}_{\text{BF}} \neq 0$. Assuming that the interaction allows the hopping of the fermion to the excited state only when there is a boson on the lowest single-particle level, the only non-zero two-body matrix elements are $v_{ij\alpha\beta} = v_{0010} =$

$v_{0001} \in \mathbb{R}$. Write $\hat{H}$ using the basis where $\hat{H}_0 = \hat{T} + \hat{V}_0$ is diagonal, i.e., the single-particle states

$$\{|10;20\rangle, |01;20\rangle, |10;11\rangle, |01;11\rangle, |10;02\rangle, |01;02\rangle\},$$

with energies

$$\{E_0, E_1, E_2, E_3, E_4, E_5\},$$

respectively, fulfilling the condition:

$$E_0 < E_1 < \cdots < E_5.$$

Find the ground state. (Advice: set $E_0 = 0$ to simplify the calculations). We have used the occupation number notation for the kets: $|n_0^f n_1^f; n_0^b n_1^b\rangle = |n_0^f\rangle_f \otimes |n_1^f\rangle_f \otimes |n_0^b\rangle_b \otimes |n_1^b\rangle_b$.

(a) Bosons and fermions are not identical particles (i.e., they are distinguishable), and our Hamiltonian must reflect this. To start with, the single-particle states of fermions and bosons need not be identical, and therefore we need to choose different single-particle indices. We will choose Latin indices for bosons and Greek indices for fermions, and we will use $\hat{b}, \hat{b}^\dagger$ for bosonic operators, and $\hat{c}, \hat{c}^\dagger$ for fermionic operators. The one-body operators are:

$$\hat{T} = \sum_{ij} t_{ij}^b \hat{b}_i^\dagger \hat{b}_j + \sum_{\alpha\beta} t_{\alpha\beta}^f \hat{c}_\alpha^\dagger \hat{c}_\beta,$$

$$\hat{V}_0 = \sum_{ij} v_{ij}^b \hat{b}_i^\dagger \hat{b}_j + \sum_{\alpha\beta} v_{\alpha\beta}^f \hat{c}_\alpha^\dagger \hat{c}_\beta,$$

where we have used the superindex b to refer to bosonic matrix elements, and f for the fermionic ones. The two-body potential reads

$$\hat{V}_{BF} = \sum_{i\alpha j\beta} v_{i\alpha j\beta} \hat{b}_i^\dagger \hat{c}_\alpha^\dagger \hat{b}_j \hat{c}_\beta = \sum_{ij\alpha\beta} v_{ij\alpha\beta} \hat{b}_i^\dagger \hat{b}_j \hat{c}_\alpha^\dagger \hat{c}_\beta.$$

Here we have used the fact that the canonical operators of distinct particle species always commute, $[\hat{b}_i, \hat{c}_j] = [\hat{b}_i, \hat{c}_\alpha^\dagger] = [\hat{b}_i^\dagger, \hat{c}_\alpha] = [\hat{b}_i^\dagger, \hat{c}_\alpha^\dagger] = 0$. Because of that, notice that, contrary to the purely fermionic case, the ordering of the destruction operators need not be reversed with respect to that of the indices (see chapter 2).

(b) We have that, both for fermions and for bosons, the single-particle energies of the (single-particle) state $|0\rangle$ are lower than that of the state $|1\rangle$. Besides, we have that $\hat{H}_0$ has only one-body operators (no inter-particle interaction), and therefore there is no mixing of single-particle states. With this information, the ground

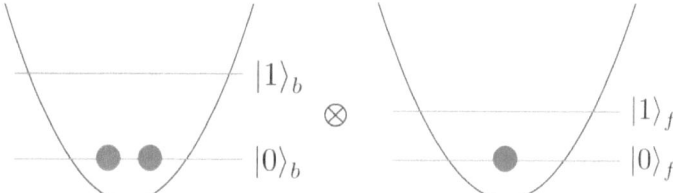

Figure 13.1 Left: potential energy well with two energy levels and two bosons in the single-particle bosonic ground state. Right: similar to the left but for a single fermion. The tensor product of the two wells represents the lowest energy level of the many-body state. Both the bosonic and fermionic ground-state energies are set to 0, for convenience.

state must be:

$$|GS\rangle = |0\rangle_f \otimes (|0\rangle_b)^2 := |10;20\rangle = \frac{1}{\sqrt{2}}\hat{c}_0^\dagger \hat{b}_0^\dagger \hat{b}_0^\dagger |vac\rangle.$$

The factor $1/\sqrt{2}$ comes from the totally-symmetric character of the bosonic part of the state. This information can be summarized in Fig. 13.1.

We will compute E_0 by computing the following expectation value.

$$\langle \hat{H} \rangle = \langle 10;20| \sum_{ij}(t_{ij}^b + v_{ij}^b)\hat{b}_i^\dagger \hat{b}_j + \sum_{\alpha\beta}(t_{\alpha\beta}^f + v_{\alpha\beta}^f)\hat{c}_\alpha^\dagger \hat{c}_\beta |10;20\rangle.$$

Using the rules of how creation and annihilation operators act on Fock states, the result is:

$$\langle \hat{H} \rangle = \sum_{ij}(t_{ij}^b + v_{ij}^b)(\sqrt{2})^2 \delta_{j0}\delta_{i0} + \sum_{\alpha\beta}(t_{\alpha\beta}^f + v_{\alpha\beta}^f)(\sqrt{1})^2 \delta_{\beta0}\delta_{\alpha0}$$

$$= 2(t_{00}^b + v_{00}^b) + (t_{00}^f + v_{00}^f) = 2e_{0,b} + e_{0,f}.$$

As expected for a non-interacting system, the total energy is the sum of the single-particle energies, so it could have been guessed without any computation.

(c) Let us start by computing the matrix elements of $\hat{V}_{BF}$. First, looking at the term with $v_{ij\alpha\beta} = v_{0010}$, the only non-zero matrix elements are:

$$\langle 01;20| \hat{V}_{BF} |10;20\rangle = \frac{1}{2}v_{0010} \langle vac |\hat{b}_0 \hat{b}_0 \hat{b}_0^\dagger \hat{b}_0 \hat{b}_0^\dagger \hat{b}_0^\dagger| vac \rangle \langle vac |\hat{c}_1 \hat{c}_1^\dagger \hat{c}_0 \hat{c}_0^\dagger| vac \rangle$$

$$= \frac{1}{2}v_{0010} 4 \langle vac |\hat{b}_0 \hat{b}_0 \hat{b}_0^\dagger \hat{b}_0 \hat{b}_0^\dagger \hat{b}_0^\dagger| vac \rangle \langle vac |\hat{c}_1 \hat{c}_1^\dagger \hat{c}_0 \hat{c}_0^\dagger| vac \rangle$$

$$= 2v_{0010},$$

which is the associated energy for the transition of the fermion between the ground state and the excited state when there are two bosons in their single-particle state.

$$\langle 01;11|\hat{V}_{\mathrm{BF}}|10;11\rangle =v_{0010}\,\langle \mathrm{vac}\,|\hat{b}_1\hat{b}_0\hat{b}_0^\dagger\hat{b}_0\hat{b}_0^\dagger\hat{b}_1^\dagger|\,\mathrm{vac}\rangle\,\langle \mathrm{vac}\,|\hat{c}_1\hat{c}_1^\dagger\hat{c}_0\hat{c}_0^\dagger|\,\mathrm{vac}\rangle$$

$$=v_{0010}\,\langle \mathrm{vac}\,|\hat{b}_1\hat{b}_0\hat{b}_0^\dagger\hat{b}_0\hat{b}_0^\dagger\hat{b}_1^\dagger|\,\mathrm{vac}\rangle\,\langle \mathrm{vac}\,|\hat{c}_1\hat{c}_1^\dagger\hat{c}_0\hat{c}_0^\dagger|\,\mathrm{vac}\rangle$$

$$=v_{0010}\,,$$

which is the associated energy for the transition of the fermion between the ground state and the excited state when there is one boson in the single-particle state. The value is half of the other element because there are fewer bosons allowing the transition.

The matrix elements stemming from $v_{ij\alpha\beta}=v_{0001}$ are necessarily the adjoints of those with $v_{ij\alpha\beta}=v_{0010}$, as $\hat{H}$ must be Hermitian.

To diagonalize $\hat{H}$ it is convenient to choose the following basis ordering:

$$\{|10;02\rangle,|01;02\rangle,|10;20\rangle,|01;20\rangle,|10;11\rangle,|01;11\rangle\}\,.$$

Now, $\hat{H}$ is block-diagonal,

$$\hat{H}=\begin{pmatrix} E_4 & 0 & 0 & 0 & 0 & 0 \\ 0 & E_5 & 0 & 0 & 0 & 0 \\ 0 & 0 & E_0 & 2v_{0010} & 0 & 0 \\ 0 & 0 & 2v_{0010} & E_1 & 0 & 0 \\ 0 & 0 & 0 & 0 & E_2 & v_{0010} \\ 0 & 0 & 0 & 0 & v_{0010} & E_3 \end{pmatrix},$$

and we can observe that the interaction only couples 1 fermion transition, and that the first two states are not affected by this interaction.

To diagonalise this matrix, we can use the fact that the determinant of a block-diagonal matrix is the product of the determinants of the blocks:

$$\det\left(\hat{H}-\lambda\hat{I}\right)=(E_4-\lambda)(E_5-\lambda)\begin{vmatrix} E_0-\lambda & 2v_{0010} \\ 2v_{0010} & E_1-\lambda \end{vmatrix}\begin{vmatrix} E_2-\lambda & v_{0010} \\ v_{0010} & E_3-\lambda \end{vmatrix}.$$

The 2×2 determinants are easily computed:

$$\begin{vmatrix} E_0-\lambda & 2v_{0010} \\ 2v_{0010} & E_1-\lambda \end{vmatrix}=(E_0-\lambda)(E_1-\lambda)-4v_{0010}^2$$

$$=\lambda^2-\lambda\,(E_0+E_1)-4v_{0010}^2+E_0E_1\,,$$

$$\begin{vmatrix} E_2-\lambda & v_{0010} \\ v_{0010} & E_3-\lambda \end{vmatrix}=(E_2-\lambda)(E_3-\lambda)-v_{0010}^2$$

$$=\lambda^2-\lambda\,(E_2+E_3)-v_{0010}^2+E_2E_3\,.$$

The solutions of setting these polynomials to zero are:

$$\lambda_\pm = \frac{1}{2}\left(E_0 + E_1 \pm \sqrt{(E_0 + E_1)^2 + (4v_{0010})^2 - 4E_0 E_1}\right)$$

and

$$\lambda'_\pm = \frac{1}{2}\left(E_2 + E_3 \pm \sqrt{(E_2 + E_3)^2 + v_{0010}^2 - 4E_2 E_3}\right)$$

respectively. Then, the overall expression for $\det\left(\hat{H} - \lambda \hat{I}\right)$ is:

$$\det\left(\hat{H} - \lambda \hat{I}\right) = -\left(E_4 - \lambda\right)\left(E_5 - \lambda\right)\left(\lambda_+ - \lambda\right)\left(\lambda_- - \lambda\right)\left(\lambda'_+ - \lambda\right)\left(\lambda'_- - \lambda\right).$$

Since we have $E_0 < E_1 < \cdots < E_5$ and we are looking for the ground state, from the equation above, we deduce that the ground state must either have energy E_4, λ_- or λ'_-. If we rewrite E_4, λ_- and λ'_- as a function of the single-particle energies $e_k^{f,b} = t_k^{f,b} + v_k^{f,b}$, setting $e_0^f = e_0^b = 0$:

$$E_4 = 2e_1^b,$$

$$\lambda_- = \frac{1}{2}\left(e_1^f - \sqrt{(e_1^f)^2 + (4v_{0010})^2}\right),$$

$$\lambda'_- = \frac{1}{2}\left(e_1^f + 2e_1^b - \sqrt{(e_1^f)^2 + v_{0010}^2}\right).$$

We see that $\lambda_- < h_1^f < 2e_1^b = E_4$, and $\lambda'_- - \lambda_- \geq e_1^b > 0$. Therefore, the ground state will have energy λ_-.

Since the Hamiltonian is block-diagonal, what could have been a system of 6 coupled equations is now a set of 4 independent systems (one for each block). We now need to find the eigenvector with this eigenvalue, for which we will work only in the subspace associated to this eigenvalue, i.e., the subspace spanned by $\{|10; 20\rangle, |01; 20\rangle\}$.

$$\hat{H}\begin{pmatrix}\gamma \\ \delta\end{pmatrix} = \begin{pmatrix}\gamma E_0 + 2\delta v_{0010} \\ 2\gamma v_{0010} + \delta E_1\end{pmatrix} = \lambda_-\begin{pmatrix}\gamma \\ \delta\end{pmatrix}.$$

Using again the single-particle energies,

$$\left.\begin{array}{l}2\delta v_{0010} = \lambda_-\gamma \\ 2\gamma v_{0010} + \delta e_1^f = \lambda_-\delta\end{array}\right\}.$$

If we now write $\delta(\gamma)$ using the first equation and insert it into the second equation, we obtain:

$$\gamma\left(2v_{0010} + e_1^f\frac{\lambda_-}{2v_{0010}} - \frac{\lambda_-^2}{2v_{0010}}\right) = 0.$$

The solution $\gamma = 0$ is not admissible as it leads to a null 8-dimensional vector, which is the trivial solution. However, we can check that the condition in parentheses is always satisfied, and therefore there is no constraint on γ. The ground state is therefore:

$$|\mathrm{GS}\rangle = |\lambda_-\rangle = \gamma\left(|10;20\rangle + \frac{\lambda_-}{2v_{0010}}|01;20\rangle\right),$$

up to a normalization factor.

This result is somewhat expected: the ground state of the interacting system is a superposition of the ground state of the non-interacting system and the state with an excited fermion, since the interaction allows the fermionic $|0\rangle^f \to |1\rangle^f$ single-particle transition (and the opposite one). In the limit $v_{0010} \to \pm\infty$, it is easy to see that $|\mathrm{GS}\rangle = \gamma(|10;20\rangle \mp |01;20\rangle)$. Physically, this means that if we increase the interaction strength, the lowest-energy configuration will be an equal superposition of $|10;20\rangle$ and $|01;20\rangle$.

Finally, the limit $v_{0010} \to 0$ is also interesting. One can see that the ground state is then $|\mathrm{GS}\rangle = |10;20\rangle$. This was expected, since having no interaction should return the same result as in the previous section, where $\hat{V}_{\mathrm{BF}} = 0$ from the start.

A Second-quantization formalism

The natural language to describe indistinguishable particles in quantum mechanics is the so-called second quantization approach [9, 11, 3, 6, 5, 1]. Rather than working in a Hilbert space with a fixed number of particles, $\mathscr{H}(N)$, in second quantization we let go of the idea of a fixed number of particles, and form a Hilbert space that is large enough to accommodate an undetermined number of particles. This is achieved by employing the so-called *Fock space*,

$$\mathscr{F} = \bigoplus_{N=0}^{\infty} \mathscr{H}(N).$$

This Fock space contains a vacuum state, $|0\rangle$, with no particles; a complete set of single-particle states, $|\alpha\rangle$, with $\alpha = 1, 2, 3, \ldots$; a complete set of two-particle states, $|\alpha\beta\rangle$; a complete set of three-particle states, $|\alpha\beta\gamma\rangle$; and so on. These complete sets of many-particle states contain only vectors of the correct permutation symmetry, which can either be fermionic or bosonic.

The key tools to exploit the extension into Fock space are the creation, $\hat{a}^\dagger$, and destruction, $\hat{a}$, operators. These act as mappings between many-particle Hilbert spaces of different particle numbers,

$$\hat{a}_\alpha^\dagger : \mathscr{H}(N-1) \to \mathscr{H}(N),$$
$$\hat{a}_\alpha : \mathscr{H}(N) \to \mathscr{H}(N-1).$$

Let us first discuss the creation operator, and we shall derive the properties of the annihilation operator from it. We also restrict the discussion to fermions in the first instance, although the extension to bosons is relatively straightforward and discussed in detail in several literature sources [2, 15, 7]. Acting on top of the vacuum state, with no particles, the operator $\hat{a}_\alpha^\dagger$ creates a particle in the state α,

$$\hat{a}_\alpha^\dagger |0\rangle = |\alpha\rangle. \tag{A.0.1}$$

Acting on top of a one-particle state $|\beta\rangle$, the same operator creates a two-particle state,

$$\hat{a}_\alpha^\dagger |\beta\rangle = |\alpha\beta\rangle. \tag{A.0.2}$$

It is useful to demand this state to be antisymmetric from the outset. For this to occur, we must have

$$\hat{a}_\alpha^\dagger \hat{a}_\beta^\dagger |0\rangle = |\alpha\beta\rangle = -|\beta\alpha\rangle = -\hat{a}_\beta^\dagger \hat{a}_\alpha^\dagger |0\rangle.$$

We can impose this condition through the anticommutation relation

$$\hat{a}_\alpha^\dagger \hat{a}_\beta^\dagger + \hat{a}_\beta^\dagger \hat{a}_\alpha^\dagger = 0 \Rightarrow \{\hat{a}_\alpha^\dagger, \hat{a}_\beta^\dagger\} = 0. \qquad (A.0.3)$$

Taking adjoints of this relation, one can easily prove that

$$\{\hat{a}_\alpha, \hat{a}_\beta\} = 0. \qquad (A.0.4)$$

Adding a third particle to a two-body state immediately yields an anti-symmetrized three-body state. It is important to note the ordering of the states on the antisymmetrized ket. Every time we create a state, we do it on the left-hand side of the ket:

$$\hat{a}_\alpha^\dagger |\beta\gamma\rangle = |\alpha\beta\gamma\rangle . \qquad (A.0.5)$$

Some additional algebra immediately shows that fermionic creation and annihilation operators also fulfill a fundamental anticommutation relation,

$$\{\hat{a}_\alpha, \hat{a}_\beta^\dagger\} = \delta_{\alpha\beta} . \qquad (A.0.6)$$

For bosons, the relevant relations between creation and destruction operators are the following commutation relations:

$$[\hat{a}_\alpha, \hat{a}_\beta^\dagger] = \delta_{\alpha\beta} , \qquad (A.0.7)$$

$$[\hat{a}_\alpha, \hat{a}_\beta] = 0, \qquad (A.0.8)$$

$$[\hat{a}_\alpha^\dagger, \hat{a}_\beta^\dagger] = 0. \qquad (A.0.9)$$

We have seen that the creation and annihilation operators allow us to define practically many-body states. This in itself would not be useful if we could not also transform operators into their Fock space counterparts. Let us first consider operators that act over one-body properties. Think, for instance, about the kinetic energy of an N−body system of identical fermions (and hence with the same mass m),

$$\hat{T} = \sum_{i=1}^{N} \frac{\hat{p}_i^2}{2m} = \sum_{i=1}^{N} \hat{t}_i. \qquad (A.0.10)$$

The single-particle operators $\hat{t}_i$ act on one particle at a time. Similar expressions hold for other one-body properties like the total momentum, or the total external potential energy of a system of N particles. We use $\hat{t}_i$ as a generic one-body operator in the following. Since it acts on individual particles, $\hat{t}_i$ has the matrix elements

$$t_{\alpha\beta} = (\alpha|\hat{t}_i|\beta) . \qquad (A.0.11)$$

A key point of this expression is that it does not depend on the particle index i. The single-particle states $|\alpha\rangle$ are predefined, and the operator $\hat{t}_i$ is exactly the same for each particle— so the corresponding single-particle matrix elements do not know to which particle they need to be applied to.

One can construct a second quantization version of the very same operator. For operators of a one-body nature, the expression reads:

$$\hat{T} = \sum_{\alpha\beta} t_{\alpha\beta} \hat{a}_\alpha^\dagger \hat{a}_\beta. \tag{A.0.12}$$

Compared to the original expression, this operator does not explicitly depend on the number of particles since it acts and lives in Fock space. It is only when $\hat{T}$ is applied to an antisymmetric $N-$body ket that the dependence over N is restored.

One-body operators are not all there are, though. In an interacting system, a key two-body operator is the interaction operator, which provides the potential between particles i and j,

$$\hat{V} = \sum_{ij} \hat{V}_{ij} = \frac{1}{2} \sum_{i \neq j} \hat{V}_{ij}. \tag{A.0.13}$$

The second-quantized version of this operator is given by the expression,

$$\hat{V} = \frac{1}{2} \sum_{\alpha\beta\delta\gamma} v_{\alpha\beta\gamma\delta} \hat{a}_\alpha^\dagger \hat{a}_\beta^\dagger \hat{a}_\delta \hat{a}_\gamma, \tag{A.0.14}$$

where it is important to stress that the order of the last two destruction operators is inverted with respect to the matrix element,

$$v_{\alpha\beta\gamma\delta} = \left(\alpha\beta | \hat{V}_{ij} | \gamma\delta \right). \tag{A.0.15}$$

Here and in Eq. (A.0.11) we employ a parenthesis notation $|\cdot)$ in the matrix elements obtained from specific single-particle states, without considering explicitly antisymmetry. Fully symmetric or antisymmetric kets, like those in Eq. (A.0.2), are instead represented by $|\cdot\rangle$.

B Wick's theorem

In the context of quantum many-body physics, one often encounters the need to evaluate expectation values of operators and their matrix elements. In second quantization, we denote the matrix element $O_{\vec{\alpha},\vec{\beta}}$ of a generic N-body operator $\hat{O}$ as:

$$O_{\vec{\alpha},\vec{\beta}} = \left\langle \vec{\alpha} \left| \hat{O} \right| \vec{\beta} \right\rangle = \left\langle \alpha_1 \ldots \alpha_N \left| \hat{O} \right| \beta_1 \ldots \beta_N \right\rangle.$$

For the sake of this example, we introduce the vector notation $|\vec{\alpha}\rangle = |\alpha_1 \ldots \alpha_N\rangle$. Let us now assume that $\hat{O}$ is of the form $\hat{a}_\gamma^\dagger \hat{a}_\delta$ for some indices γ, δ. The matrix element then reads

$$O_{\vec{\alpha},\vec{\beta}} = \left\langle 0 \left| \hat{a}_{\alpha_N} \ldots \hat{a}_{\alpha_1} \hat{a}_\gamma^\dagger \hat{a}_\delta \hat{a}_{\beta_1}^\dagger \ldots \hat{a}_{\beta_N}^\dagger \right| 0 \right\rangle.$$

We now have an expectation value on the vacuum of a product of creation and annihilation operators. In general, in second quantization, we can always express any matrix element, expectation value, or overlap between two states as a vacuum expectation value of a product of operators.

In order to carry out actual calculations, one generally wishes to rearrange such products of operators so that all creation operators lie to the left of all annihilation operators. Whenever a product is thusly arranged, we shall say that it is in *normal order*. Why is this a good idea? Precisely because if a product $\hat{A}_1 \hat{A}_2 \ldots \hat{A}_N$ is in normal order form, then:

$$\langle 0 | \hat{A}_1 \hat{A}_2 \ldots \hat{A}_N | 0 \rangle = 0,$$

where we have used upper-case letters to denote fully general operators, i.e., that can be creation or annihilation operators. Let us see how to write any product of operators as a sum of normal-ordered terms. Take as an example the following product of 3 operators:

$$\hat{a}_\alpha \hat{a}_\beta^\dagger \hat{a}_\gamma^\dagger = \left(-\hat{a}_\beta^\dagger \hat{a}_\alpha + \delta_{\alpha\beta} \right) \hat{a}_\gamma^\dagger = \delta_{\alpha\beta} \hat{a}_\gamma^\dagger - \hat{a}_\beta^\dagger \left(-\hat{a}_\gamma^\dagger \hat{a}_\alpha + \delta_{\alpha\beta} \right)$$
$$= \hat{a}_\beta^\dagger \hat{a}_\gamma^\dagger \hat{a}_\alpha - \delta_{\alpha\beta} \hat{a}_\gamma^\dagger - \delta_{\alpha\gamma} \hat{a}_\beta^\dagger.$$

The right-hand side in this last line is formed by a term, $\hat{a}_\beta^\dagger \hat{a}_\gamma^\dagger \hat{a}_\alpha$, which is in normal order, minus two additional terms involving Kronecker δ's on single-particle indices.

The process to arrive at normal order products of creation and annihilation operators becomes quite tedious as the number of operators in the product increases. Gian Carlo Wick (1909 - 1992) suggested an easier way to evaluate such products in his 1950 article entitled *Evaluation of the Collision Matrix* [14]. Before we announce this theorem, we need to introduce some definitions.

Definition (normal ordering operator). Consider the product of operators Π that contains both creation and annihilation operators. The *normal ordering operator*

$\mathcal{N}$ acting on Π is defined as:

$$\boxed{\mathcal{N}(\Pi) \equiv (\pm 1)^P \Pi(\text{creation} \times \text{annihilation})}, \qquad (\text{B.0.1})$$

where the $+$ sign corresponds to the case where all operators in Π are bosonic, and the $-$ sign, to the case where they are all fermionic. P is the sign of the permutation that takes all creation operators to the left of all annihilation operators. One can think of the above equation as rearranging the original product Π in normal order by using the fermionic (bosonic) anticommutation (commutation) relations, and then ignoring the terms with Kronecker deltas.

Notice that the normal ordering operator outcome is unique up to the reordering of creation (or annihilation) operators within themselves. In fact, let us assume that a product is already in normal order, and now we wish to exchange two creation (or annihilation) operators. If these operators are bosonic, they will commute, and therefore we end up with a reordering of the same product. If they are fermionic, since we are effectively ignoring the Kronecker deltas, we are introducing an extra minus sign at most. However, this will increase the parity of the overall permutation by one, and therefore the overall factor $(-1)^P$ will counteract the change of sign, leaving us with the same value for the reordering of the initial product.

Definition (contraction). We define the *contraction* of two operators $\hat{A}, \hat{B}$ as:

$$\boxed{\overline{\hat{A}\hat{B}} \equiv \hat{A}\hat{B} - \mathcal{N}(\hat{A}\hat{B})}. \qquad (\text{B.0.2})$$

Because $\mathcal{N}(\hat{A}\hat{B})$ is only a rearrangement of $\hat{A}\hat{B}$, the contraction of two operators is actually an (anti)commutation relation, and therefore a $c-$number.

The usefulness of this definition lies partly in the fact that, since the vacuum expectation of a normal-ordered product vanishes, and since a contraction is just a $c-$number, we have:

$$\overline{\hat{A}\hat{B}} = \langle 0|\overline{\hat{A}\hat{B}}|0\rangle = \langle 0|\hat{A}\hat{B}|0\rangle.$$

Definition. We define the contraction of two operators $\hat{A}, \hat{B}$ which are separated, within a product, by a chain of n operators, as:

$$\boxed{\overline{\hat{A}\hat{A}_1 \ldots \hat{A}_n \hat{B}} \equiv (\pm 1)^n \overline{\hat{A}\hat{B}}\hat{A}_1 \ldots \hat{A}_n}, \qquad (\text{B.0.3})$$

where the $+$ sign corresponds to the bosonic case, and the $-$ sign, to the fermionic case. n can also be seen as the number of operator transpositions which takes the contracted pair to the left of the product.

Let us now see how these definitions can help us in rearranging operator products. We start by looking at some simple products of fermionic operators and manually expressing them in normal order. The product of two canonical operators is

$$\hat{a}_\alpha \hat{a}_\beta^\dagger = -\hat{a}_\beta^\dagger \hat{a}_\alpha + \delta_{\alpha\beta},$$

and by the definition of a contraction, Eq. (B.0.2), the following must also be true:

$$\hat{a}_\alpha \hat{a}_\beta^\dagger = \mathcal{N}(\hat{a}_\alpha \hat{a}_\beta^\dagger) + \overparen{\hat{a}_\alpha \hat{a}_\beta^\dagger}.$$

Let us now look into products with three elements:

$$\hat{a}_\alpha \hat{a}_\beta^\dagger \hat{a}_\gamma = \left(-\hat{a}_\beta^\dagger \hat{a}_\alpha + \delta_{\alpha\beta}\right)\hat{a}_\gamma = -\hat{a}_\beta^\dagger \hat{a}_\alpha \hat{a}_\gamma + \delta_{\alpha\beta}\hat{a}_\gamma, \tag{B.0.4}$$

$$\hat{a}_\alpha \hat{a}_\beta \hat{a}_\gamma^\dagger = \hat{a}_\alpha\left(-\hat{a}_\gamma^\dagger \hat{a}_\beta + \delta_{\beta\gamma}\right) = \delta_{\beta\gamma}\hat{a}_\alpha - \left(-\hat{a}_\gamma^\dagger \hat{a}_\alpha + \delta_{\alpha\gamma}\right)\hat{a}_\beta$$

$$= \hat{a}_\gamma^\dagger \hat{a}_\alpha \hat{a}_\beta + \delta_{\beta\gamma}\hat{a}_\alpha - \delta_{\alpha\gamma}\hat{a}_\beta, \tag{B.0.5}$$

$$\hat{a}_\alpha \hat{a}_\beta^\dagger \hat{a}_\gamma^\dagger = \hat{a}_\beta^\dagger \hat{a}_\gamma^\dagger \hat{a}_\alpha - \delta_{\alpha\beta}\hat{a}_\gamma^\dagger - \delta_{\alpha\gamma}\hat{a}_\beta^\dagger, \tag{B.0.6}$$

$$\hat{a}_\alpha^\dagger \hat{a}_\beta \hat{a}_\gamma^\dagger = \hat{a}_\alpha^\dagger\left(-\hat{a}_\gamma^\dagger \hat{a}_\beta + \delta_{\beta\delta}\right) = -\hat{a}_\alpha^\dagger \hat{a}_\gamma^\dagger \hat{a}_\beta + \delta_{\beta\gamma}\hat{a}_\alpha^\dagger. \tag{B.0.7}$$

Finally, we provide an example of a 4-operator product:

$$\hat{a}_\alpha \hat{a}_\beta \hat{a}_\gamma^\dagger \hat{a}_\delta^\dagger = \hat{a}_\alpha\left(-\hat{a}_\gamma^\dagger \hat{a}_\beta + \delta_{\beta\gamma}\right)\hat{a}_\delta^\dagger = \delta_{\beta\gamma}\hat{a}_\alpha \hat{a}_\delta^\dagger - \left(-\hat{a}_\gamma^\dagger \hat{a}_\alpha + \delta_{\alpha\gamma}\right)\hat{a}_\beta \hat{a}_\delta^\dagger$$

$$= \delta_{\beta\gamma}\hat{a}_\alpha \hat{a}_\delta^\dagger - \delta_{\alpha\gamma}\hat{a}_\beta \hat{a}_\delta^\dagger + \hat{a}_\gamma^\dagger \hat{a}_\alpha\left(-\hat{a}_\delta^\dagger \hat{a}_\beta + \delta_{\beta\delta}\right)$$

$$= \delta_{\beta\gamma}\left(-\hat{a}_\delta^\dagger \hat{a}_\alpha + \delta_{\alpha\delta}\right) - \delta_{\alpha\gamma}\left(-\hat{a}_\delta^\dagger \hat{a}_\beta + \delta_{\beta\delta}\right) + \delta_{\beta\delta}\hat{a}_\gamma^\dagger \hat{a}_\alpha$$

$$\quad - \hat{a}_\gamma^\dagger\left(-\hat{a}_\delta^\dagger \hat{a}_\alpha + \delta_{\alpha\delta}\right)\hat{a}_\beta$$

$$= \delta_{\beta\gamma}\delta_{\alpha\delta} - \delta_{\alpha\gamma}\delta_{\beta\delta} - \delta_{\beta\gamma}\hat{a}_\delta^\dagger \hat{a}_\alpha + \delta_{\alpha\gamma}\hat{a}_\delta^\dagger \hat{a}_\beta + \delta_{\beta\delta}\hat{a}_\gamma^\dagger \hat{a}_\alpha - \delta_{\alpha\delta}\hat{a}_\gamma^\dagger \hat{a}_\beta$$

$$\quad + \hat{a}_\gamma^\dagger \hat{a}_\delta^\dagger \hat{a}_\gamma \hat{a}_\beta. \tag{B.0.8}$$

There is a clear pattern in all the above expressions. A normal-ordered product always appears as one of the terms in the r.h.s., and then we have a sum of terms containing deltas and normal-ordered products with fewer operators. First, we can think of the Kronecker deltas as the contraction of two operators (see Problem 3.1). Second, by the definition of normal ordering it is immediate that we can freely take any multiplicative factor in and out of the normal ordering operation, so we can actually express all r.h.s. terms as normal-ordered products with contractions. Take for instance the last term in Eq. (B.0.4):

$$\delta_{\alpha\beta}\hat{a}_\gamma = \overparen{\hat{a}_\alpha \hat{a}_\beta^\dagger}\mathcal{N}(\hat{a}_\gamma) = \mathcal{N}(\overparen{\hat{a}_\alpha \hat{a}_\beta^\dagger}\hat{a}_\gamma).$$

This can be expressed as a normal-ordered and contracted term. The same applies to all the other terms on the r.h.s. of Eqs. (B.0.4)-(B.0.8) that have Kronecker deltas. In some cases, however, we may need to be careful with the fermionic sign. Consider the third term in Eq. (B.0.5),

$$\delta_{\alpha\gamma}\hat{a}_\beta = \overparen{\hat{a}_\alpha \hat{a}_\gamma^\dagger}\mathcal{N}(\hat{a}_\beta) = -\mathcal{N}(\overparen{\hat{a}_\alpha \hat{a}_\beta \hat{a}_\gamma^\dagger}),$$

where we require a $-$ sign by virtue of Eq. (B.0.3).

Taking these ideas into consideration, we can re-express all the products above in a way that involves only sums (no negative signs) of normal-ordered and contracted terms:

$$\hat{a}_\alpha \hat{a}_\beta^\dagger = \mathcal{N}(\hat{a}_\alpha \hat{a}_\beta^\dagger) + \mathcal{N}(\overline{\hat{a}_\alpha \hat{a}_\beta^\dagger})$$

$$\hat{a}_\alpha \hat{a}_\beta^\dagger \hat{a}_\gamma = \mathcal{N}(\hat{a}_\alpha \hat{a}_\beta^\dagger \hat{a}_\gamma) + \mathcal{N}(\overline{\hat{a}_\alpha \hat{a}_\beta^\dagger} \hat{a}_\gamma)$$

$$\hat{a}_\alpha \hat{a}_\beta \hat{a}_\gamma^\dagger = \mathcal{N}(\hat{a}_\alpha \hat{a}_\beta \hat{a}_\gamma^\dagger) + \mathcal{N}(\hat{a}_\alpha \overline{\hat{a}_\beta \hat{a}_\gamma^\dagger}) + \mathcal{N}(\overline{\hat{a}_\alpha \hat{a}_\beta \hat{a}_\gamma^\dagger})$$

$$\hat{a}_\alpha \hat{a}_\beta^\dagger \hat{a}_\gamma^\dagger = \mathcal{N}(\hat{a}_\alpha \hat{a}_\beta^\dagger \hat{a}_\gamma^\dagger) + \mathcal{N}(\overline{\hat{a}_\alpha \hat{a}_\beta^\dagger} \hat{a}_\gamma^\dagger) + \mathcal{N}(\overline{\hat{a}_\alpha \hat{a}_\beta^\dagger \hat{a}_\gamma^\dagger})$$

$$\hat{a}_\alpha^\dagger \hat{a}_\beta \hat{a}_\gamma^\dagger = \mathcal{N}(\hat{a}_\alpha^\dagger \hat{a}_\beta \hat{a}_\gamma^\dagger) + \mathcal{N}(\hat{a}_\alpha^\dagger \overline{\hat{a}_\beta \hat{a}_\gamma^\dagger})$$

$$\hat{a}_\alpha \hat{a}_\beta \hat{a}_\gamma^\dagger \hat{a}_\delta^\dagger = \mathcal{N}(\hat{a}_\alpha \hat{a}_\beta \hat{a}_\gamma^\dagger \hat{a}_\delta^\dagger)$$

$$+ \mathcal{N}(\hat{a}_\alpha \hat{a}_\beta \overline{\hat{a}_\gamma^\dagger \hat{a}_\delta^\dagger}) + \mathcal{N}(\overline{\hat{a}_\alpha \hat{a}_\beta \hat{a}_\gamma^\dagger} \hat{a}_\delta^\dagger) + \mathcal{N}(\hat{a}_\alpha \overline{\hat{a}_\beta \hat{a}_\gamma^\dagger} \hat{a}_\delta^\dagger) + \mathcal{N}(\overline{\hat{a}_\alpha \hat{a}_\beta \hat{a}_\gamma^\dagger \hat{a}_\delta^\dagger})$$

$$+ \mathcal{N}(\overline{\hat{a}_\alpha \hat{a}_\beta \overline{\hat{a}_\gamma^\dagger} \hat{a}_\delta^\dagger}) + \mathcal{N}(\overline{\hat{a}_\alpha \hat{a}_\beta \hat{a}_\gamma^\dagger \hat{a}_\delta^\dagger}).$$

Notice that, in each of these expressions, we could have added normal-ordered terms involving contractions of the sort: $\overline{\hat{a}_\alpha \hat{a}_\beta}$, $\overline{\hat{a}_\alpha^\dagger \hat{a}_\beta^\dagger}$, $\overline{\hat{a}_\alpha^\dagger \hat{a}_\beta}$, as they are all null (see Problem 3.1). This, in turn, is an indication that we might be able to express any product of operators as a sum of normal-ordered products with different numbers of contractions. This result, obtained here with rather hand-wavy arguments, can be rigorously proven to be true. This is in fact the well-known *Wick's Theorem*:

Theorem (Wick's Theorem):

$$\hat{A}_1 \ldots \hat{A}_n = \mathcal{N}(\hat{A}_1 \ldots \hat{A}_n)$$

$$+ \sum_{(ij)} \mathcal{N}(\hat{A}_1 \ldots \overline{\hat{A}_i \ldots \hat{A}_j} \ldots \hat{A}_n)$$

$$+ \sum_{(ij)(kl)} \mathcal{N}(\hat{A}_1 \ldots \overline{\hat{A}_i \ldots \hat{A}_k \ldots \hat{A}_j} \ldots \hat{A}_l \ldots \hat{A}_n)$$

$$\vdots$$

$$+ \text{sum over fully-contracted products}. \qquad (B.0.9)$$

The first sum runs over all single contractions; the second sum runs over all combinations of double contractions; the last term runs over all full contractions, i.e., where all operators are contracted. The proof of this theorem is somewhat long and will not be given here. The interested reader can find it in G. C. Wick's 1950 paper [14] and other books [3, 5, 7, 9].

The theorem is of particular use in the following situations: whenever we have a vacuum expectation value, it is clear that all normal-ordered terms will cancel. By Wick's theorem, the only terms that remain are the fully-contracted contributions from the last line, such that:

$$\langle 0 | \hat{A}_1 \ldots \hat{A}_N | 0 \rangle = \sum_{\text{all pair combinations}} (\pm 1)^P \times \text{fully-contracted products} \,, \quad \text{(B.0.10)}$$

where P is the number of transpositions needed to take all contracted pairs together. When writing vacuum expectation values of fully-contracted products, it is common practice to omit the normal-ordering symbol. In fact, the result of performing a full contraction is a c-number, so the normal-ordering does not play a role anyway.

There is one last result which greatly simplifies the calculation of any expectation value using Wick's theorem:

Theorem (Sign Rule and Line Crossings): *the factor* $(-1)^P$ *appearing in Eq. (B.0.10) equals the number of contraction line crossings.*

This allows us to directly compute the fermionic phase associated to normal ordering from the contractions themselves.

References

1. A. Altland and B. D. Simons. *Condensed Matter Field Theory*. Cambridge University Press, 2 edition, 2010.

2. L. E. Ballentine. *Quantum Mechanics: A Modern Development*. World Scientific, 1998.

3. J.-P. Blaizot and G. Ripka. *Quantum theory of finite systems*. MIT Press, Cambridge, Massachusetts, 1986.

4. P. Coleman. *Intoduction to Many-Body Physics*. Cambridge University Press, 2015.

5. W. H. Dickhoff and D. van Neck. *Many-body theory exposed!: propagator description of quantum mechanics in many-body systems*. World Scientific, first edition, 2005.

6. A. L. Fetter and J. D. Walecka. *Quantum Theory of Many-Particle Systems*. Dover, NY, 2003.

7. E. K. U. Gross, E. Runge, and O. Heinonen. *Many-Particle Theory*. Taylor & Francis, 1991.

8. X.-W. Guan, M. T. Batchelor, and C. Lee. Fermi Gases in One Dimension: From Bethe Ansatz to Experiments. *Rev. Mod. Phys.*, 85:1633–1691, Nov 2013.

9. G. D. Mahan. *Many-particle Physics*. Plenum Press, NY, Second edition, 1990.

10. N. Navon, S. Nascimbène, F. Chevy, and C. Salomon. The Equation of State of a Low-Temperature Fermi Gas with Tunable Interactions. *Science*, 328(5979):729–732, 2010.

11. J. W. Negele and H. Orland. *Quantum Many-Particle Systems*. Addison-Wesley, 1987.

12. Gianluca Stefanucci and Robert van Leeuwen. *Nonequilibrium Many-Body Theory of Quantum Systems: A Modern Introduction*. Cambridge University Press, first edition, 2013.

13. A. N. Wenz, G. Zürn, S. Murmann, I. Brouzos, T. Lompe, and S. Jochim. From Few to Many: Observing the Formation of a Fermi Sea One Atom at a Time. *Science*, 342(6157):457–460, 2013.

14. G. C. Wick. The Evaluation of the Collision Matrix. *Phys. Rev.*, 80:268–272, Oct 1950.

15. V. Zelevinsky. *Quantum Physics: Volume 2*. Wiley, 2010.

Index

antisymmetrization operator, 3, 5

BCS ansatz, 136, 141
BCS theory, 135

canonical transformation, 22
closure relation, 41
Compton profile, 107
creation operator, 151

density operator, 18
destruction operator, 151
distribution function, 93
dynamic structure function, 102

Fermi distribution, 75
Fermi energy, 40, 63
Fermi momentum, 52
field operator, 17, 19
Fourier transform, 42
free particle propagator, 38

Green's function, 37

Hartree-Fock energy, 112, 116
Hartree-Fock method, 111
Heisenberg representation, 20

Koltun sum-rule, 47

linear response function, 101

one-body density matrix, 123
one-body operator, 14, 29
operator contraction, 156

particle number operator, 14, 18, 136
permutation symmetry, 3

second quantization, 13
sequential condition, 57, 94, 126
Slater determinant, 111
Slater function, 57, 94, 124
spectral function, 40, 45, 47

spin operator, 23
symmetrization operator, 3, 5
symmetrization postulate, 3

three-body distribution function, 126
two-body density matrix, 128
two-body distribution function, 54, 57, 94, 126
two-body operator, 14, 31

wavefunction, 3
Wick's theorem, 27